Liaison Functions 2021

AA009022

Held at the 2021 AIChE Annual Meeting

Boston, Massachusetts, USA and Online
7 - 11 November and 15 - 19 November 2021

ISBN: 978-1-7138-5712-9

Printed from e-media with permission by:

Curran Associates, Inc.
57 Morehouse Lane
Red Hook, NY 12571

Some format issues inherent in the e-media version may also appear in this print version.

Printed with permission by Curran Associates, Inc. (2022)

AIChE
120 Wall Street, FL 23
New York, NY 10005-4020

Phone: (800) 242-4363
Fax: (203) 775-5177

www.aiche.org

Additional copies of this publication are available from:

Curran Associates, Inc.
57 Morehouse Lane
Red Hook, NY 12571 USA
Phone: 845-758-0400
Fax: 845-758-2633
Email: curran@proceedings.com
Web: www.proceedings.com

TABLE OF CONTENTS

635621 New High-Performance Hydrocracking Catalyst Platform

Abstract:

Honeywell UOP has a rich history of innovation in hydrocracking catalyst development and we have consistently renewed our portfolio by introducing new hydrocracking catalysts with improved performance. Recently, using fundamental science and state-of-the-art characterization techniques, we have developed a family of hydrocracking catalysts which has superior performance over our existing catalyst portfolio. These catalysts were developed on a newly improved commercial base tailored with desirable pore structure. The metals system was also optimized to achieve optimum interactions between the active metals and the catalyst support. In the presentation, details of the development of this new catalyst platform along with advanced characterization results and catalytic performance will be discussed.

635620 First-Principles Modeling of the Effect of Surface Modification Strategies on Hydrogenation Catalysis

Abstract:

Common methods to tune the performance of heterogenous transition metal catalysts include combining two or more metals to form an alloy, synthesizing shape- or size-specific nanoparticles, and utilizing a catalyst support to impart bifunctionality. In addition, but less widely applied, the surface of a transition metal catalyst can be tailored through controlled monolayer deposition of organic or inorganic surface modifiers. Here, two recent examples of surface modification of Pd, leading to enhanced hydrogenation activity, will be discussed. In the first example, a partial monolayer TiO_2 overcoat deposited on Pd via atomic layer deposition is found to modulate the relative binding strength of surface intermediates in naphthalene hydrogenation leading to an increase in turnover frequency relative to the uncoated Pd catalyst.[1] In a second example, the addition of organic phosphonic acid (PA) modifiers to Pd is shown to enhance ring hydrogenation of furfural alcohol through hydrogen-bonding interactions between the surface bound intermediates and the PA modifiers.[2] For both of the examples discussed, the unexpected hydrogenation activity enhancement resulting from addition of the surface modifier to Pd is revealed through first-principles atomic-scale modeling of the underlying surface chemistry.

1. W. W. McNeary, S. A. Tacey, G. Lahti, D. R. Conklin, K. A. Unocic, E. C. D. Tan, E. C. Wegener, T. E. Erden, S. Moulton, C. Gump, J. Burger, M. B. Griffin, C. A. Farberow, M. J. Watson, L. Tuxworth, K. M. Van Allsburg, A. A. Dameron, K. Buechler, D. R Vardon, *ACS Catalysis* (2021) – in press.
2. P. Coan, C. A. Farberow, M. B. Griffin, J. W. Medlin, *ACS Catalysis* 11, 3730 (2021).

635615 Building Catalysts, Reactors and Engineers for a Sustainable Future

Abstract:

Our planet must address pressing global issues such as climate change, food and water quality, and growing energy demand. Chemical engineering, including reactor design and catalysis development, are crucial to efficient production of fuels and chemical precursors and removal of toxins and pollutants from water and food sources. My research at Lawrence Livermore National Laboratory focuses on pioneering potential solutions to these challenges by combining advanced manufacturing (AM) and electrochemical engineering. Together, these approaches can be used to design and control catalyst structures, fluid dynamics, and mass transport phenomena from the sub-micron through the centimeter length scales; furthermore, AM allows for the exploration of novel reactor designs that would be otherwise unattainable through other manufacturing approaches. In this work, I demonstrate how we produced the first 3D-printed vapor-fed reactor for electrochemical CO_2 reduction, achieved remarkable selectivity and production of multi-carbon products such as ethylene and ethanol, and shifted the paradigm on how we create next generation of chemical reactor design. Finally, I will share how my nonprofit, the Feaster Foundation, focuses on encouraging and supporting underserved students who are passionate about helping others; this underscores my firm belief that the development of brilliant engineers is just as important as the development of incredible reactors.

This work was performed under the auspices of the U.S. Department of Energy by Lawrence Livermore National Laboratory under Contract DE-AC52-07NA27344.

635616 Metal-Organic Frameworks (MOFs) As Catalysts and Catalyst Precursors for Oxidation and Reduction Reactions

Abstract:

Metal-organic frameworks (MOFs) are porous materials that are formed by nodes of metal ions or metal oxide clusters linked by organic ligands; the modularity of these components results in a large set of diverse pore networks useful in various applications. Here, we investigate MOFs as catalysts and catalyst precursors for oxidation, reduction, and dehydrogenation reactions. First, we probe the kinetic mechanism and recyclability of MIL-101 in its Cr and Fe forms for styrene oxidation by hydrogen peroxide. Styrene consumption turnover rates and product distributions are relatively constant across all reaction cycles for MIL-101(Fe), whereas MIL-101(Cr) resulted in increased rates and a different product distribution after the first cycle; the presence of solvated Cr(III) complexes was observed via UV-Vis, indicating that Cr leaching dominates cycling inefficiencies. Second, we utilize MOFs as catalysts precursors that result in unique nanomaterials upon high temperature treatment or under an applied potential. The efficacy of these MOF-derived catalysts, in monometallic Fe and Cu forms, was tested for propane dehydrogenation and CO_2 electrocatalytic reduction, respectively. The effect of external applied driving force on the phase and/or structure of the catalysts, as determined from various characterization techniques, was evaluated along with their catalytic performances. The work shown here provides examples of using MOFs as catalysts and as precursors to synthesize unique, stable catalyst structures with desired rates and selectivities, which we can leverage to optimize green methods for producing valuable organic molecules. Further, this work combines expertise that Michele garnered from working in four different heterogeneous catalysis labs, beginning as a sophomore undergraduate student. The varying scientific techniques and mentoring philosophies have shaped her nascent independent research group at Princeton University.

635609 Probing Dynamic Materials and Systems for Electrocatalytic Processes

Abstract:

Research in the Seitz lab aims to revolutionize catalytic processes at the foundation of our energy and chemical industries that are currently major drivers of climate change. Progress in this area is impeded by knowledge gaps of complex reaction mechanisms, intermediates, and catalysts needed to transform these industries. We seek to fill these gaps using an innovative combination of electrochemistry, spectroscopy, and materials science towards the development of catalyst materials and reactor systems with enhanced activity, selectivity, and stability. Electrochemical processes are a promising alternative to conventional thermochemical combustion processes as they operate at atmospheric temperature and pressure, produce fewer pollutants, and provide a mechanism for storage and conversion of vital renewable electricity sources.

We use controlled material syntheses and advanced spectroscopy techniques to monitor dynamic behavior or catalysts in response to reaction conditions. Notably, we have developed several iridium-based perovskite materials to establish electronic structure effects associated with systematic changes in composition, crystallinity, and strain. We use these materials to elucidate trends in structural reorganization and degradation mechanisms induced by reaction operating conditions, exemplified with water oxidation in acidic conditions, for which active and stable catalyst development has been a longstanding challenge. We also investigate the relationship between reactor bulk and catalyst local reaction environments to determine changes in reaction mechanism and product selectivity as a function of bulk electrolyte pH, concentration, cation identity, and operating potential. Local pH probes and kinetic isotope effect studies provide insights to behavior of carbon-based catalysts for the oxygen reduction reaction in systematically varied reaction environments. Understanding how to control complex electrochemical reaction mechanisms via catalyst and environment tuning is critical for reducing energy consumption and optimizing reaction selectivity towards desired product molecules that serve as critical building blocks for fuels and products that we rely on every day.

634962 "Black Lives Matter in Science, Engineering and Medicine" and "Success is What You Leave Behind"

Abstract:

In two talks, Dr. Laurencin will focus on the two areas for which he received the Hoover Medal: his work in social justice, and his work in mentoring.

In his first talk, delivered when he received the Herbert W. Nickens Award, he will discuss addressing racism as being key in moving forward to create a more equitable environment. Dr. Laurencin describes the critical need to shift from diversity, inclusion, and equity (D.I.E.) to inclusion, diversity, equity, anti-racism, and learning (I.D.E.A.L.)

In his second talk, he will discuss the 14 principles of success he has taught his students. The principles are a part of his forthcoming biography published by Elsevier, entitled "Success is What You Leave Behind".

636455 Use of Environmental Sustainability Principles to Help Solve Environmental Problems of Concern Today

Abstract:

In 1987, the concept of sustainable development was first put forward in the Bruntland Report which was prepared for the World Commission on Environment and Development. The links between inequality, poverty, and environmental degradation were recognized; the report sought a way for a fairer world with less damage to the environment that supports communities. The report defined sustainable development as "Development that meets the needs of the present without compromising the ability of future generations to meet their own needs". As a consequence, environmental sustainability ensures that in meeting our needs for water, food, and shelter as well as engaging in activities that make our lives enjoyable, we don't damage the environment, deplete our natural resources, and protect global ecosystems. Important aspects tied in with address: (1) Behavioral economics; (2) Intergenerational equity; and (3) What to do with the developing world. As engineers, we will increasingly need to be considering these concepts.

Recent research activities aimed toward environmental sustainability have involved: condensate recovery in air handlers; improved building energy conservation; use of green roofs combined with cistern collection and rain gardens for storm water mitigation on campus buildings; modeling of pollutants fate and transport in the Nile River tributaries; use of electrochemical sensors in storm drains to characterize storm water runoff in urban roadways; use of vertical gardening systems and community gardens to address urban food deserts; lead contamination in drinking water systems physical/chemical treatment of hazardous wastes, particularly in the area of sonication (acoustic cavitation) and photocatalysts to treat organic contaminants; and waste plastics in the marine environment. Highlights of several of these research topics will be described in the context of environmental sustainability.

636472 Advancing Methodologies for Sustainability assessment of transition to Low Carbon and Circular Economy via integration of process systems engineering and macroeconomics

Abstract:

Decarbonization and implementing Circular Economy are two major goals for sustainable development to mitigate the challenges of climate change and increasing anthropogenic wastes. Among numerous pathways proposed for this transition, it is important to evaluate the systems level economic and environmental impacts of these pathways for selection of the most sustainable pathways. Industrial Ecology (IE) is a field of study that focuses on developing methodologies and computational tools to assess large scale impact of these interventions. One key method in IE utilizes the macroeconomic Input-Output model for evaluating the impact of planned transitions such as shifts to renewable energy, bio-based production etc. However, the existing techniques do not efficiently integrate process engineering and thermodynamics knowledge in sustainability assessments, thus limiting the automation in evaluation of new technologies under development at scale. Further, lack of engineering integration and reliance on only monetary based model introduces uncertainty in analysis due to price variation and obsolete models along with missing feedback in considering sustainability criteria in design. In this talk, I will present the advancement in systems level sustainability assessment techniques by addressing these limitations through integrating process systems engineering and thermodynamics with macroeconomic Input-Output framework. Specifically, I will present novel methodologies and computational tools that will enable fast, and reproducible quantitative evaluation of systems level environmental and economic impacts of adopting emerging technologies for decarbonization and circular economy pathways.

634961 Reclaiming the Engineering in the Minds of the Public:The Unheralded, Underappreciated, and Misunderstood Method that Built Our Modern World

Abstract:

Naively the public assumes the products of engineers arise from the scientific method, as reflected in an old joke among engineers about the relationship of science and engineering: “if it’s a success, then it’s a scientific miracle, if a disaster, then an engineering failure.” This joke highlights that successful technologies are invisible: The hallmark of good engineering is invisibility — we rarely think of our furnace, or a jet’s engine, or the purity of a pharmaceutical because the methods to manufacture all these have been honed to perfection. This, though, also hides the creative work of engineers because the public assumes the secret of engineering lies in the mastery of arcane realms of knowledge — sophisticated calculus and powerful computing science implemented by a dispassionate, almost mechanical person — yet the power of engineers to change the world lies in their method, a method used long before sophisticated mathematics and computers. This talk lifts the veil to show, in all its glory, the engineering method, which, once understood, highlights the creativity of engineers, demonstrates their work is the pinnacle of human reasoning, and lays a foundation about how to think about technology — how to decide its proper use and aid it in fulfilling its promise. Using rich examples, this talk strips bare the tools often confused for the engineering method – scientific knowledge, mathematical manipulation – to expose what lies at the heart of the method: a surprisingly simple notion called a “rule of thumb.”

634959 Directed Evolution of New Adeno-Associated Viral Vectors for Clinical Gene Therapy

Abstract:

Gene therapy – the delivery of genetic material to the cells of a patient for therapeutic benefit – has been increasingly successful in human clinical trials over the past decade, and there are currently five FDA-approved gene therapies. The most successful gene delivery vehicles, or vectors, are based on adeno-associated viruses (AAV); however, vectors based on natural versions of AAV face a number of delivery barriers that limit their efficacy and will thus preclude the extension of these successes to the majority of human diseases. These delivery limitations arise since the parent viruses upon which these vectors are based were not evolved by nature for our convenience to use as human medicine. Unfortunately, due to the highly complex mechanisms of virus-host interactions, there is currently insufficient mechanistic knowledge to enable rational design to be sufficiently successful in creating new vectors. As an alternative, however, we developed the concept of using directed evolution to engineer highly optimized variants of AAV for a broad range of cell and tissue targets. Directed evolution involves the iterative genetic diversification of a biomolecule to create a gene pool and functional selection to isolate variants with optimal properties. Using this approach, we have engineered AAV variants with greatly improved delivery efficiency to multiple organs including the retina; lung, and muscle; targeted delivery to specific cell types; and the capacity to evade immune responses. Our novel AAV variants are currently used in 5 human clinical trials involving delivery to the retina and heart, and additional products will enter into the clinic this year.

In parallel, the advent of genome editing technologies such as the CRISPR/Cas9 system raise the possibility of using gene delivery not only for gene replacement but for repair or knockout of endogenous genes. We have thus been combining engineered AAVs with CRISPR/Cas9 for a range of applications. The integration of these new technologies – AAV delivery or genome editing machinery – can enable a broad range of basic and therapeutic applications.

638225 Direct ink write 3D printing of high solids loading bimodal distributions of particles

Abstract:

As additive manufacturing expands as a processing technology for structurally customizable materials, there is increasing interest in printing with high particle contents. For suspensions with particle contents of over 50 vol% there are significant formulation and processing challenges due to increased interparticle friction and suspension complexity. We focus on suspensions with bimodal particle distributions and two common binder systems, a high molecular weight polymer in a solvent that solidifies via solvent evaporation and a monomer mixture that cures via UV irradiation. We examine the interplay between formulation and processing and show that the formulation effects are particularly important at optimal printing parameters, but that they are overcome by processing-related defects at sub-optimal parameters. By understanding the processing and formulation effects specific to direct ink write additive manufacturing of high solids suspensions, new customized inks can be designed for a wide range of applications, including construction, energetics and ceramics.

638226 Modeling phase formation on catalyst surfaces: Coke formation andsuppression in hydrocarbon environments

Abstract:

We develop a simulation toolset employing density functional theory (DFT) in conjunction with grand canonical Monte Carlo (GCMC) to study coke formation on Fe-based catalysts during propane dehydrogenation (PDH). As expected, pure Fe surfaces develop stable graphitic coke structures and rapidly deactivate. We find that coke formation is markedly less favorable on Fe3C and FeS surfaces. Fe-Al alloys display varying degrees of coke resistance, depending on their composition, suggesting that they can be optimized for coke resistance under PDH conditions. Electronic structure analyses show that both electron-withdrawing effects (on Fe3C and FeS) and electron-donating effects (on Fe-Al alloys) destabilize adsorbed carbon. On the alloy surfaces, a geometric effect also isolates Fe sites and disrupts the formation of graphitic carbon networks. This work demonstrates the utility of GCMC for studying the formation of disordered phases on catalyst surfaces and provides insights for improving the coke resistance of Fe-based PDH catalysts.

638227 Modular Gas-to-Liquids Process with Membrane Steam-Methane Reformer and Fischer–Tropsch Reactive Distillation

Abstract:

Challenges related to energy and the environment, alongside risks in capital investment

decisions have sparked recent interest in process intensification research. Process

intensification has been characterized by novel equipment designs that produce dramatic process improvements including reduction in size, waste production, and energy consumption. Given the current pressing environmental and social concerns on the shale gas industry, it is imperative to design and optimize energy systems to produce solutions that are economically efficient, environmentally sustainable, and socially responsible. Toward this goal, the main research challenge is to develop an integrated energy system modeling framework that can systematically generate optimal designs and operational strategies, and comprehensively improve multiple sustainability criteria. This work highlights opportunities to transform gas-to-liquids plants to compact, safe, energy efficient, and sustainable processes by using intensified units such as membrane reformers for feed handling and reactive distillation to take advantage of reaction heat todrive separation and enhance yield.

638228 Building a queryable nanotherapeutic database to advance nanomedicine translation to the clinic

Abstract:

Many nanomedicine-focused labs produce 100s to 1000s, or more, of individual nanoparticle formulations across the lab lifetime. Ideally, when developing and optimizing our nanoformulations, we rapidly identify a formulation most likely to translate to the clinic for a specific indication. Each formulation represents an experimental data point intended to contribute to our growing understanding and implementation of nanomedicine for the treatment of disease. Yet, the time it takes to develop a nanotherapeutic to the point of translation often results in testing of many formulations by a number of researchers over many years across a range of experimental conditions and in multiple biological models. In addition, it is important for translation to understand the formulation-dependent interactions between nanomedicine platforms and the complex biological environment in which they are applied. To investigate dynamic nanoparticle-biology interactions with a range of nano-based delivery systems, nanomedicine labs may find it useful to develop a database management schema that can organize and integrate key experimental variables and factors that might influence the effectiveness of nanotherapeutics in the target setting.

In this talk, we explore the development of a queryable database management system using our labs 6 years of nanotherapeutic formulations for treating brain disease. Database management systems already commonly exist in clinical medical applications. For example, The Brain Database is a clinical neuroanatomy database developed over three decades ago. However, pre-clinical data management schemas have different complexities than clinical applications. Pre-clinical databases need management schemas that logically connect experimental methodologies, which can be highly repetitive, vary in quality, are prone to rapid iteration and evolution, and often research lab or facility-specific. There is limited or non-existent literature on effective database management for nanoformulations used preclinically, but similar approaches for utilizing databases exist in computational cell biology for managing data-rich biological cell process data. A successful standardized data management schema for pre-clinical nanotherapeutic experimental pipelines will have three main effects: (1) increased sustainability for data organization and storage, (2) increased insight into related nanotherapeutic parameters and their biological interactions, and (3) increased searchability of specific results or methods.

To show the efficacy of a well-designed pre-clinical database management schema for nanoformulation data, we create a database using PostgreSQL that performs three functions: (1) Obtains the nanoparticle formulation variables that produce specific nanoparticle sizes and charges; (2) Connects nanoformulations to all related experimental elements, including nanoformulation methodology, biological specimen prep, and biological specimen characterization; and (3) Defines a standard lab protocol for regularly updating the database as new nanoformulation experiments are carried out. We show we can query our nanoformulations in this database to obtain nanoformulation batches with specific physicochemical properties or to identify relationships between nanoformulations and specific biological outcomes, such as impact on brain cell viability. This talk highlights the proof-of-concept that our experimental data can benefit from a well-structured database management schema. It also lays the foundation to develop standard practices for discussing, planning, and participating in better data management. These practices will lead to better data utilization and data longevity, and greater reproducibility and interpretability of datasets generated from independent researchers.

A standard data management schema for experimental nanoformulation experiments decreases the need for repetitive experimentation, connects nanoformulation variables with biological outcome in an interpretable and efficient way, and enables science-informed querying of any standardized variables. The overall benefits of a standardized database contribute to efficient experimentation and cross-experimental insight which will ultimately improve pre-clinical to clinical translation of nanoformulations.

638229 Structure of Nanoparticle-Polyelectrolyte Complexes: Effects of Polyelectrolyte Characteristics and Charge Ratio

Abstract:

We present structural and turbidimetric characterizations of aqueous dispersions of oppositely charged nanoparticles and polyelectrolytes that undergo complexation and flocculation to form particle-rich agglomerates over a wide range of polyelectrolyte and nanoparticle concentrations with varying polyelectrolyte sizes and persistence lengths. Compaction of nanoparticles in the polyelectrolyte-nanoparticle complexes, signified by decreasing inter-nanoparticle spacing until they are brought in near-contact, is observed upon increasing screening of interparticle repulsion arising from adsorption of polyelectrolytes on the nanoparticle surfaces. We identify the ratio of charge born by the polyelectrolytes and the nanoparticles as a key parameter dictating the compaction of nanoparticles into complexes as well as densification of complexes into agglomerates. Moreover, the emergence of turbidity in solution is shown to correlate with the initiation of nanoparticle compaction by polyelectrolytes but failing to describe further structural evolution of the complexes, emphasizing the need for alternate characterization techniques.

638230 Scalable 3D Printed Lattices for Pressure Control in Fluid Applications

Abstract:

Additive manufacturing affords precise control over geometries with high degrees of complexity and pre-defined structure. Lattices are one class of additive-only structures which have great potential in directing transport phenomena because they are highly ordered, scalable, and modular. However, a comprehensive description of how these structures scale and interact in heterogeneous systems is still undetermined. To advance this aim, we designed cubic and Kelvin lattices at two sub-5 mm length scales and compared published correlations to the experimental pressure gradient in pipes ranging from 12-52 mm diameter. We further investigated all combinations of the four lattices to evaluate segmented combinatorial behavior. The results suggest that a single correlation can describe pressure behavior for different lattice geometries and scales. Furthermore, combining lattice systems in series has a complex effect that is sensitive to part geometry. Together, these developments support the promise for tailored, modular lattice systems at laboratory scales and beyond.

636125 Exploiting Redox for Connecting Biology to Electronics – Controlling Behavior via Electrogenetics

Abstract:

The Internet of Things has been a truly transformative reality - of great successes, opportunities and challenges. Big data and electronic information transfer, however, typically stops at the interface with biology, it's aqueous nature largely devoid of free electron transfer. By recognizing that biological redox active molecules are a biological equivalent of an electron-carrying wire, we have developed biological surrogates for electronic devices, including a biological redox capacitor that enable bi-directional "electron" flow. We have also turned to synthetic biology to provide a means to sample, interpret and report on biological information contained in molecular communications circuitry. We have based much of our work on the cell-cell molecular communication system known as bacterial quorum sensing, which mediates the phenotypic transition from single cells to a collective. That is, we and many others have participated in elucidating the molecular and genetic regulation of bacterial quorum sensing (QS) and have "rewired" its regulatory components so as to enable "eavesdropping" on bacterial crosstalk. We have also developed synthetic genetic circuits that exploit intracellular redox "nodes" to enable electronic actuation of gene expression in QS and other systems. Using simple reconstructions, one can apply voltage on an electrode and directly actuate genetic responses and associated phenotypes. We have turned to information theory to help guide genetic multiplexing for maximum information transfer. We have exploited these circuits to develop systems that autonomously control subpopulations within bacterial co-cultures. This presentation will introduce the concepts of molecular communication that are enabled by integrating relatively simple concepts in synthetic biology with biofabrication. Our presentation will show how engineered cells represent a versatile means for mediating the molecular "signatures" commonly found in complex environments, or in other words, they are conveyors of molecular communication.

636126 Synthetic Biology: Life Redesigned

Abstract:

Synthetic biology is bringing together engineers, physicists and biologists to model, design and construct biological circuits out of proteins, genes and other bits of DNA, and to use these circuits to rewire and reprogram organisms. These re-engineered organisms are going to change our lives in the coming years, leading to cheaper drugs, rapid diagnostic tests, and synthetic probiotics to treat infections and a range of complex diseases. In this talk, we highlight recent efforts to create synthetic gene networks and programmable cells, and discuss a variety of synthetic biology applications in biotechnology and biomedicine.

634960 Viruses, Immunity, and Vaccines

Abstract:

Infectious disease-causing pathogens have plagued humanity since antiquity, and the COVID-19 pandemic has been a vivid reminder of this perpetual existential threat. Vaccination has saved more lives than any other medical procedure, and indeed, effective vaccines now promise to end the COVID-19 pandemic. However, we do not have effective vaccines against rapidly mutating viruses, such as HIV; nor do we have a universal vaccine against seasonal variants of influenza. The ability to develop effective vaccines that protect us from highly mutable viruses will help create a more pandemic-resilient world. In this lecture, I will describe how by bringing together approaches from engineering and the life and physical sciences, progress is being made to address this challenge. Specifically, I will describe approaches that aim to define the mutational vulnerabilities of mutable viruses, and design vaccines that elicit immune responses that can target these vulnerabilities. The application of these approaches toward the development of effective HIV and influenza vaccines will be discussed.

634700 The Synergy between Experiment and Theory in Catalysis Research

Abstract:

Recent advances in theoretical methods and accelerated computing powers have open doors to easily access and use quantum mechanical calculations in researching heterogeneous catalysis. Such methods, in turn, have enabled to understand the nature of active centers in solid catalysts and the detailed reaction mechanisms, which were often difficult to assess solely from experiments. Indeed, the development of density functional theory (DFT) methods have allowed to make a breakthrough in catalysis research, and now, it is a common practice in the field to collaborate between experiment and theory to solve difficult problems. Yet, rigorously connecting experiment and theory has remained as a challenge, including a strategy to find appropriate catalyst models that accurately capture the nature of active sites in working conditions. In this work, we demonstrate how to rigorously combine kinetic, isotopic, spectroscopic, and transient experiments with DFT calculations to reveal the mechanistic details of HCOOH dehydration on TiO_2 catalysts and their acid-base properties that determine their reactivities.

In-situ infrared spectra, combined with transient experiments, demonstrate that all Ti_{5c}-O_{2c} acid-base site pairs in TiO_2 samples are saturated with bidentate formates (*HCOO*) during catalytic turnovers (0.1-4 kPa HCOOH; < 500 K), requiring theory to model TiO_2 surfaces fully saturated *HCOO* species. In doing so, we show that HCOOH adsorbs molecularly (HCOOH*) on top of the *HCOO* adlayer to form a second adlayer through interactions of HCOOH* with the H-atom in the *HCOO* adlayer. These HCOOH* species undergo dehydration via a transition state that is stabilized by interactions with a Ti_{5c} center, which is made available only by the momentary formation of HCOOH* from a vicinal *HCOO*. Such mechanistic interpretation agrees well with dehydration kinetics and kinetic isotopic effects and explains the 10-times lower dehydration rates on rutile TiO_2 than of anatase TiO_2; stronger Ti_{5c} acid centers in rutile bind *HCOO* species much more strongly, requiring larger energy to "shove" such *HCOO* species for HCOOH* dehydration. The results of this work, in turn, illustrate how diverse experimental tools and theoretical methods must collaborate to provide definitive mechanistic conclusions and how these methods must accept and rigorously consider the crowded nature of surfaces during practical catalysis.

634741 Photoelectrochemistry of Semiconducting Polymers: Engineering Electronic Properties and Charge Carrier Dynamics

Abstract:

A liquid junction photoelectrochemical (PEC) cell is an attractive mean to generate clean and renewable energy (i.e., hydrogen and hydrocarbons) from solar energy. Despite the extensive research in PEC systems using inorganic semiconductors, it remains challenging to develop efficient, stable, and inexpensive PEC systems. Semiconducting polymers are potential candidates satisfying the criteria that materials consist of earth abundant materials and the polymer-based devices can be prepared by low cost and large-scale processing techniques such as inkjet and roll-to-roll printings. Moreover, tunable electronic properties of semiconducting polymers offer an opportunity to develop efficient, stable, and inexpensive PEC systems. Doping of semiconductors is an essential process to control electronic properties of the materials, thus making feasible applications in PEC cells and improving performance of the cells. However, there is a lack of fundamental study of doping to tailor electronic properties of semiconducting polymers for PEC applications. In this study, we discuss doping of p-type semiconducting polymers and their photoactivities in PEC cells. First, irreversible doping is desired for stable PEC operations since doped semiconducting polymers should maintain their electronic properties without unwanted de-doping during the operations. We found that the irreversible doping can be achieved by doping of bulky and/or hydrophilic dopant species into the semiconducting polymer films, leading to significant increases in photoactivities. Secondly, transient photocurrent measurements were conducted to investigate charge carrier dynamics of semiconducting polymers in PEC cells. The detailed understanding on improved photoactivities of doped semiconducting polymers would pave a road to further develop the high-performance polymer-based PEC systems.

634498 Dynamics of Wetting Transitions in Complex Systems

Abstract:

Studies on wetting transition, or how the meniscus of a liquid droplet suspended over the asperities on a rough solid surface ("Cassie-Baxter state") transitions to a Wenzel state where the liquid now occupies the space between the asperities, have been extensive and provided much insight into scientific and industrial applications such as plant leaves, skin, fabrics, as well as the development and design of everyday products such as skin creams, paints, and epidermal medicines where liquids are applied on rough and often porous surfaces. While most of these studies are focused on the equilibrium states of the liquids, they seldom discuss the mechanism with and rates at which the wetting transition occurs. In addition, most of those studies on the equilibrium states of wetting transition discuss one kind of liquid droplet placed on a rough solid surface while both of them are surrounded by air (hereby called solid-liquid-vapor systems, or SLV systems). While we encounter situations where two liquids are involved, such as skin creams, as often as we do with SLV systems, they did not attain much attention. Therefore, this talk presents the rate of wetting transition for both SLV systems and SL_1L_2 systems (solid-liquid 1-liquid 2)

First, for SLV systems, the dynamics of the wetting transitions are presented both experimentally and theoretically with water-air-silicon wafer systems where the Wenzel state is always the thermodynamically favorable state, while a temporary metastable Cassie-Baxter state can also exist to determine the variables that control the rates of such transitions. Theses variables are identified to affect the rate of wetting transition: (i) the intrinsic contact angle, (ii) the concentration of dissolved air in the bulk water phase, (iii) the liquid volatility that determines the rate of capillary condensation inside the cavities, and (iv) the presence of surfactants.

Secondly, for SL_1L_2 systems, oil-water-silicon wafer systems were used to investigate the rate of wetting transition. When two liquids are involved, one encounter oil-water mixture most frequently during dairy processes and enhanced oil recovery with low-salinity water flooding. The experiments with deionized water revealed that solubility and diffusivity were the driving factors that promoted the replacement, and the 3D confocal images confirmed the decrease in volume of hydrocarbons due to dissolution and diffusion. To exclude these factors, water was saturated with each of the seven hydrocarbons prior to experiments. For these experiments, the adhesion energy of hydrocarbons to the solid surface and the density difference between hydrocarbons and water were correlated with the rate.

634306 3D Trajectories, Diffusion, and Interactions of Single Ceria Particles on a Glass Surface and Their Removal in Post-CMP Applications

Abstract:

A very important goal of dielectric chemical mechanical planarization (CMP) is to uniformly polish a dielectric material such as silicon dioxide in the IC manufacturing processes. Ceria-based slurries have been widely employed in such a dielectric CMP process, but the abrasive particles can remain attached to the oxide films after polishing and need to be removed completely during the subsequent cleaning process. As the minimum feature size of integrated circuit elements has shrunk below 7 nm, post-CMP cleaning technologies have become significantly more challenging. There has been increasing demand to understand various phenomena that occur during cleaning.

Recently, we proposed evanescent wave (EW)-based optical detection methods integrated with video microscopy to study various chemical and mechanical phenomena such as slurry chemistry, interfacial force measurement, surface adsorption, and surface characterization/ modification during polishing and cleaning. 3D trajectories, diffusion coefficients, and interactions of ceria particles near a glass surface in the slurry at the different pH values were investigated. How these results are related to the removal of ceria particles by repeated washing with DI water adjusted to pH 10 and 12 will be discussed. Here, the flat glass surface of the lens became our model for the SiO2 film surface of interest. In situ measurement data collected during cleaning are likely to provide more important insights into the underlying mechanisms of various complex reactions that occur at the particles/brush–wafer interface. As mentioned above, there is a strong need to understand various chemical and mechanical phenomena that occur in the brush cleaning systems used to remove the ceria particles adsorbed on the SiO2 film. This talk will also introduce our most recent EW microscopy coupled with a brushed set-up to monitor the real-time visualization of the removal of single ceria particles from the SiO2 film with PVA brush scrubbing.

634307 Robust Interfacial Tension Measurement and Surfactant Formulation Evaluation Protocol for Enhanced Oil Recovery

Abstract:

Chemical enhanced oil recovery (cEOR) aims to improve the oil recovery rate of mature oil fields. The interfacial tensions (IFTs) between the aqueous phases, the oil phase, and other possible phases from the oil/water mixtures are important for designing and estimating a chemical formulation effectiveness. Furthermore, ultralow ($< 10^{-2}$ $mN \cdot m^{-1}$) IFTs are targeted to increase the capillary number and help mobilize trapped oil droplets. However, determination of reliable IFT values for surfactant/oil/brine mixtures involving commercial surfactants and crude oil have been avoided, because it is difficult to determine such values mainly due to the complexity of the mixture systems. Here, a robust protocol for determining reliable equilibrium IFT (EIFT) values for a commercial surfactant, crude oil, and synthetic reservoir brine is demonstrated to improve the conventional surfactant formulation design process for target reservoirs. The protocol contains dynamic measurement of IFT values and controlled interfacial area perturbation tests to test the stability of the steady-state IFT values.

Moreover, five types of EIFTs have been defined to establish the relation between the EIFT values of surfactant/oil/brine mixtures and laboratory-scale core flood test results: (I) the un-pre-equilibrated equilibrium IFT; (II) the un-pre-equilibrated EIFTs in the presence of rock; (III) the pre-equilibrated EIFTs in the presence of oil; (IV) the pre-equilibrated EIFT in the presence of rock and oil; and (V) the effluent EIFT. Among the five types of EIFTs, the fourth and fifth were found to be the most relevant for achieving high oil recovery performance in core flood tests, because they capture the effects of surfactant partitioning into the oil phase and the adsorption losses on the rock surface. For instance, for three surfactant formulations tested with Berea core flood experiments, the one with the lowest type-IV EIFT (~0.01 $mN \cdot m^{-1}$) had the highest oil recovery ratio (78%), and the one with the highest type IV EIFT (~0.2 $mN \cdot m^{-1}$) had the lowest oil recovery ratio (55%). The other EIFTs correlated less well with the oil recovery performance. Identifying surfactant formulations that have low or ultralow EIFTs, especially of type IV or V, are critical for screening formulations appropriate for core flood tests and target field applications, and for predicting oil recovery performance.

621594 Over Half a Century of Protecting the Environment through Various Policies, Administrations, Successes, & Failures; An Overview of 50+ Years of EPA Regulations and Standards

Abstract:

Since the establishment of the U.S. Environmental Protection Agency (EPA) on December 2nd, 1970, the Agency has implemented a vast number of regulations to benefit the American people and the environment. Over the last 51 years, the EPA and its dedicated workers have endeavored to fulfil its mission to protect human health and the environment; the Agency has sustained its critical function through 51 years and 10 different administrations (6 Republican and 4 Democratic presidents). The change of administration leads to change and priorities, and at times new or rolled back regulations. One such case is the Bush-era fuel standard increasing the fuel standards (established in 1975) to a corporate average of 35 miles per gallon by 2020, further increased by the Obama-era fuel efficiency standard of 54.5 mpg by 2025, then replaced by the Trump-era standard requiring a small 1.5% increase in average fuel economy compared to the pre-Obama standards, and then upon the inauguration of the Biden-era, auto makers themselves began to support higher fuel efficiency standards as the Administration foreshadowed increasing the standards akin to the Obama-era. This presentation aims to provide an overview of current U.S. environmental regulations, where we currently are, where we have come from, and where we are likely to go in the future.

628185 Decarbonising Residential Heating: Cost-Benefit Analysis to Support Policy Design

Abstract:

The UK's legally binding targets to achieve "net zero" greenhouse gas emissions by the year 2050 may require full decarbonisation of residential heating. This will involve deploying low-carbon heating technologies (such as heat pumps, electric storage heaters, and/or various hydrogen heating technologies) across the country, including in the 24 million homes that currently use natural-gas-based central heating systems. A key challenge of this transition is that the capital and operating costs of low-carbon heating systems are often greater than current natural-gas-based systems. In addition, the UK has a wide variety of building stock, which defies a one-size-fits-all solution.

The aim of this study is to assess the suitability of various policy instruments to promote the uptake of low-carbon heating technologies. A cost-benefit model is being developed based on housing stock data from the English House Survey, performance data from field trials, and cost estimates from the literature. The model will be used to calculate the lifetime cost and levelised cost of heat for a wide range of households and the results analysed to consider how the costs of different low-carbon heating options vary with characteristics such as building type, number of occupants, thermal efficiency, and geographical location. In addition, an uncertainty analysis will used to evaluate the impact of different fuel prices, outdoor temperatures, and personal preferences and to identify the household types likely to be impacted most by these uncertainties.

The results of this study will be presented and discussed in relation to current policies designed to promote low-carbon heating. This study will consider what level of government support will be sufficient to promote low carbon heating over the current natural-gas-based heating systems and will identify if a targeted approach based on geographic location, building type, tenure type, or other characteristics would be more appropriate.

625673 Improving the Economics of Industrial Battery Storage: A Proactive Policy and Management Approach

Abstract:

Historically, implementation of battery energy storage has been proportionally higher in commercial and residential systems than industrial systems [1]. Economic and operational constraints of industrial facilities make battery installation close to infeasible. Industrial energy usage accounts for about 25% of total energy usage in the US. As the deployment of battery storage increases, lack of implementation in the industrial sector creates a gap, bridged by adapting control and policy strategies to constraints of industrial users. Projections indicate that battery prices will continue to fall, making it even more important to find ways to implement batteries in the industrial sector [2].

This research aims to show that the utilization of electric battery storage to decrease the energy costs and peak power usage of industrial energy users can be made attainable by identifying and closing the policy gap between affordability and utility of industrial battery installations. Similar studies have been done to improve economics and management of storage in residential installations [3]. Other studies have looked at regulatory design of energy storage coupled with renewable sources such as solar and wind [4]. Though these efforts have furthered the penetration of battery storage in the residential and commercial sectors, this same penetration has not been seen in the industrial sector from lack of analysis and specificity to the needs of industry. Policies specific to large electrical users in the industrial sector have also not been explored.

Load data from actual industrial facilities is used spanning a full year of typical manufacturing operations in which the daily electrical peak is about 2200 to 2500 kW. Two models are used, a battery model utilizing the facility load data to look at battery dynamics and effects of system load, and a financial model to analyze overall cost and effect of policy on project economics. The study identifies policies that are currently in place in parts of the US or similar policies that would allow batteries to be financially viable for industrial facilities.

[1] EIA, "Battery Storage in the United States : An Update on Market Trends," Washington, DC, 2020.

[2] W. Cole and A. W. Frazier, "Cost Projections for Utility-Scale Battery Storage," 2019.

[3] M. N. Sheha and K. M. Powell, "An economic and policy case for proactive home energy management systems with photovoltaics and batteries," *Electr. J.*, vol. 32, no. 1, pp. 6–12, 2019, doi: 10.1016/j.tej.2019.01.009.

[4] O. Zinaman, T. Bowen, and A. Aznar, "An overview of behind-the-meter solar-plus-storage regulatory design: Approaches and case studies to inform international applications (Report no. NREL/TP-7A40-75283)," *NREL Rep.*, no. March, pp. 1–76, 2020.

634697 Surface Science - Nexus of Tribology, Glass Science, and Plant Biology

Abstract:

Our group has been pioneering the application of surface science principles and characterization methods to address important questions in these seemingly disparate research frontiers – tribochemistry, corrosion of silicate glasses, and plant cell walls. Being an outsider in each of these research fields, we bring radical views as well as new scientific tools to solve decades-old, highly-elusive fundamental questions that engineers and scientists equipped with traditional tools and methodologies could not solve. This talk will cover the following examples: (i) How can interfacial shear activate chemical reactions that normally do not occur? (ii) Will the subsurface structure of a glass be altered after elastic contact with a foreign object and what would be consequences of such structural changes in chemical durability? (iii) Why do plants deposit cellulose in their cell walls with different nano-to-meso scale structures?

634468 High-Performance Chiral Organic Optoelectronics Using Supramolecular Chirality

Abstract:

Chiral supramolecular nanostructures with optoelectronic functions are expected to open new opportunities in many scientific and technological fields. I will introduce fabrication of chiral supramolecular assemblies using organic semiconductors and their structure-chiroptical property relationships. I will present synthesis of chiral supramolecular biocoordination polymers and their applications in various sensors. I will also introduce a skeleton merging approach through distortion of a perylene diimide core with four fused heteroaromatics, which enables high-performance broadband circularly polarized light detections up to NIR spectral region. These findings provide guidelines for enhancing chiroptical properties using supramolecular chirality and rational molecular design toward high-performance chiral optoelectronics.

634430 Toward Single Atom Catalysis for Environmental Application

Abstract:

Various noble metals have been employed as catalysts and co-catalysts to enhance the kinetics of reactions that are critical for environmental remediation. The need to maximize the available catalytic sites per unit mass, particularly due to noble metal scarcity and high cost, has driven the development of material architectures at the nanometer scale. Recent theoretical and experimental studies have explored noble metal catalysts on the sub-nanometer, atomic scale to maximize atomic efficiency. A single atom catalyst (SAC), also called atomically dispersed catalyst, is the theoretical limit in this endeavor, and previous studies have identified additional benefits such as low coordination state and strong interaction with the surrounding substrate resulting in enhanced selective catalysis. For past few years, our group has since been exploring various SACs (Pt, Pd, Ag, and Co) anchored on different substrates (SiC, C3N4, and TiO2) that exhibit unique catalytic properties. This talk summarizes our recent studies to advance the water treatment catalysts from the nano-scale toward the single atom scale such as (1) single-atom Pt and Pd anchored to SiC for selective defluorination of perfluorooctanoic acid (PFOA) and selective dehalogenation of various water pollutants and (2) spatial separation of two co-catalysts by coordinating single atom cobalt at the void center of C_3N_4 and anchoring anthraquinone at the edges of C_3N_4 platelets to significantly enhance the catalytic synthesis of H_2O_2, a precursor chemical for advanced oxidation.

Orientation of Biomaterial based liquid crystal phases and its application

Dong Ki Yoon
Dept. of Chemistry, KAIST, Daejeon, 34141, Korea
E-mail: nandk@kaist.ac.kr

Organization of biopolymer is significant for not only an evolution of a biological system but also construction of material for well-defined nanostructures. Recently, several researchers have investigated a formation of liquid crystalline pattern of semi-flexible biopolymers such as a DNA and Sunset yellow (SSY).[1-3] However, the pattern of biopolymers was formed on restricted area that makes this intriguing patterns hard to utilize in useful applications.[4,5]

In the recent work, the 2D liquid crystalline patterns of DNA was generated when the mechanical shearing was applied on the aqueous DNA sample under the topographic confinement in drying process, in which the spontaneous shear-aligning process was a key idea to make the regularly arranged micro-paterns.[6,7] The aligned patterns were analyzed by polarized optical microscopy (POM) and atomic force microscopy (AFM). We believe the 2D micro-patterns fabricated in large area can suggest a hint to use DNA materials for organizing multiscale hierarchical structures of soft- and biomaterials.

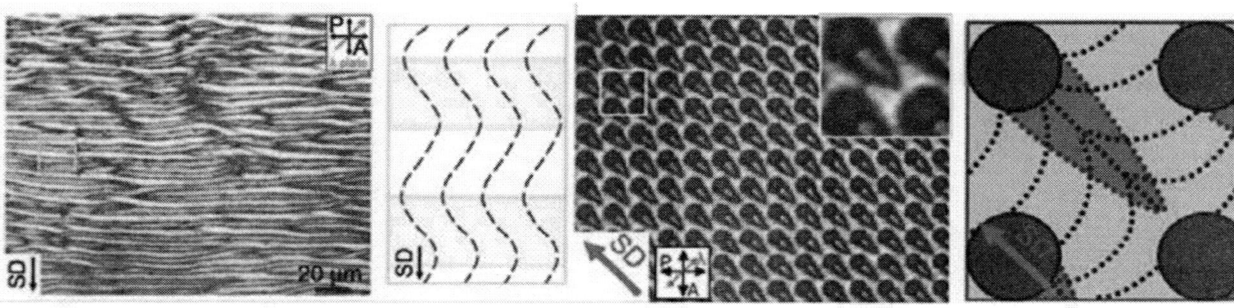

POM images of aligned zigzag and 'Ice cream cone'-like patterns of DNA material, and schematic illustrations of each color box (Black dotted lines are parallel with DNA chains).

References

1. L. Vonna et al., *Langmuir*, 21, 9635-9643 (2005).
2. I. I. Smalyukh et al., *Phy. Rev. Lett.* 96, 177801 (2006).
3. Y. J. Cha and D. K. Yoon, *Adv. Mater.* 9, 1604247 (2017)
4. Y. J. Cha et al., *ACS Appl. Mater. Inter.*, 7, 13627 (2015)
5. Y. J. Cha, D. S.Kim and D. K. Yoon, *Adv. Func. Mater.* 27, 1703790 (2017)
6. Y. J. Cha, S. M. Park, R. You, H. Kim and D. K. Yoon, *Nat. Commun.*, 10, 2512 (2019)
7. S.M. Park et al., *Small*, 16, 2002449 (2020)

635121 Engineering Vesicles Made from Recombinant Proteins Towards Synthetic Cells

Abstract:

The creation of a synthetic cell via self-assembly has been highlighted to help understand how complex biological cells come about from non-living molecules. Also, engineering synthetic vesicle platforms can provide a systematic tool to realize sophisticated cellular functions in synthetic materials. Although amphiphilic lipids and block copolymers have widely been used to construct vesicles - the essential cell structure, recombinant proteins have recently gathered great attention to developing more complex biological functions in synthetic cell platforms. We make globular protein vesicles (GPVs) by tuning the self-assembly of recombinant proteins including functionally folded globular domains. The GPV offers significant advantages not yet realized in existing protocell platforms, such as versatile integration of functional proteins with precise control of the vesicle physical properties. In this talk, I will discuss the engineering strategies to control their structural stability upon external stimuli triggers, membrane structure, and permeability of GPVs, which are critical for integrating biological functions into synthetic vesicles. Our ongoing work and future plans for making GPVs towards synthetic cells, including de novo protein synthesis, enzymatic cascade reactions, and sensing of biochemicals will also be discussed.

635122 Process-Structure-Property Relationships of Stimuli-Responsive Composite Hydrogels

Abstract:

The design and fabrication of stimuli-responsive composite hydrogels with reversible shape morphing and tunable actuation speed have been of great interest. In this talk, I will discuss our recent progress in approaches to integrate functional materials (i.e., graphene oxide, nanoclay, etc) into stimuli-responsive hydrogels (i.e., poly(N-isopropylacrylamide)). These approaches provide insights into how the presence of additives and physical interactions between additives and hydrogel networks can lead to different internal structures, and thus mechanical properties. Next, I will describe how we have leveraged these insights to precisely program the 3D shapes and control the swelling-deswelling kinetics by external stimuli. Lastly, I will also demonstrate how the different manufacturing processes can affect the micro/macro-structures and properties of composite hydrogels. Such stimuli-responsive shape morphing systems of composite hydrogels will allow us to understand their process-structure-property relationships and utilize these systems for applications ranging from soft robotics to biomedical devices.

633293 Particle Suspensions in Elastic Fluids: From Synovial Fluid Therapy to a "Swimming Rheometer"

Abstract:

Rigid or flexible particles suspended in viscoelastic fluids are ubiquitous in the food industry (e.g. pastes), industrial molding applications (all composites and 3-D printed parts), the energy industry (e.g. fracking fluids), and biological fluids (i.e. swimming of bacteria in mucous). Bill Schowalter was a pioneer in examining these suspensions particularly examining porous media flows and drop flows in elastic liquids in the 70's and 80's. The mathematical description of these suspensions is remarkably still in its infancy, but a the real breakthrough in this area has been the development of 3D computational simulations of such viscoelastic suspensions including particle motion and particle level resolution of the elastic flow fields. These simulations will allow the principles which govern the simplest flows of such suspensions, which are now generally not understood, to become elucidated in the next decade. I will describe two problems that have been recently analyzed using these computational methods – including the shear thickening of such suspensions in model synovial fluids and the design of a mechanical swimmer that is sensitive to elastic fluid rheology.

635268 Chinese American Chemical Society: Overview and 40th Anniversary Celebration

Abstract:

Founded in 1981, the Chinese American Chemical Society (CACS) is a non-profit professional society. We serve Chinese American students, practicing, and retired chemists, chemical engineers, as well as those working in related fields. Our mission is broad, providing opportunities for fellowship and recognition, support for leadership and career development, and networking for job and research opportunities.

With neither a national nor regional political affiliation, CACS also collaborates with ACS, AIChE, SAPA (Sino-American Pharmaceutical Professionals Association) and other Asian and Asian American professional societies to provide a collective voice for impact and advocacy on issues of common interest.

To celebrate CACS' 40th anniversary, we have launched a series of exciting events in 2021 to supplement our distinguished keynote banquet series at national ACS and AIChE meetings. These include special symposia in both ACS and AIChE national meetings. A global leaders' session will kick off an ACS Presidential Symposium on "Chemistry Resilience: Advanced Materials and Multidisciplinary Applications—Celebrating the 40th anniversary of CACS." Three inaugural national CACS awards will also be presented. Two sessions in 2021AIChE Annual meeting, with prominent invited speakers, aims to bring together researchers to discuss the state-of-the-art research progress and development in all aspects of energy as well as the advanced materials in energy applications.

This talk will provide an overview of CACS, the National CACS leadership team, and highlights of activities from CACS local chapters as part of our exciting 40th Anniversary symposium at the ACS Fall National Meeting and AIChE Annual Meeting.

635271 Organic Material Design for Next Generation Batteries

Abstract:

In this talk, I will talk about research we are performing related to designing and understanding of ionically-conductive polymers for battery applications. We identify critical gaps that our novel polymers can fill in realizing next generation battery materials for electrical vehicles and consumer electronics.

635272 The Chemistry of Making Semiconductor Nanowires for Artificial Photosynthesis

Abstract:

Semiconductor nanowires, by definition, typically have nanoscale cross-sectional dimensions, with lengths spanning from hundreds of nanometers to millimeters. These subwavelength structures represent a new class of semiconductor materials for investigating light generation, propagation, detection, amplification, modulation as well as energy conversion and storage. After more than two decades of research, nanowires can now be synthesized and assembled with specific compositions, heterojunctions, and architectures. This has led to a host of nanowire photonic and electronic devices. Nanowire also represents an important class of nanostructure building blocks for photovoltaics as well as direct solar-to-fuel conversion because of their high surface area, tunable bandgap, and efficient charge transport and collection. In this talk, I will present a brief history of nanowire research for the past 25 years and highlight the synthesis of nanowires using well-defined chemistry. These semiconductor nanowires are then used for artificial photosynthesis, where solar energy is converted and stored in chemical bonds in a solar driven CO_2 fixation process.

635273 A Lithium-Oxygen Battery Based on Lithium Superoxide

Abstract:

Batteries based on sodium and potassium superoxide have recently been reported. However, there have been no reports of a battery based on lithium superoxide (LiO_2), despite much recent research into the lithium-oxygen (Li-O_2) battery because of its potential high energy density. Several recent studies of Li-O_2 batteries have found evidence of LiO_2 being formed as one component of the discharge product along with lithium peroxide (Li_2O_2). In addition, theoretical calculations have indicated that some forms of LiO_2 may have a significant lifetime. The recent reports of the existence of LiO_2 as one component in the discharge product would suggest it might be formed by itself for use in a battery. However, solid LiO_2 has been difficult to synthesize in pure form because it is thermodynamically unstable with respect to disproportionation to Li_2O_2. Here we show for the first time that crystalline LiO_2 can be stabilized in a Li-O_2 battery by using a suitable cathode material, with no evidence for the presence of Li_2O_2 from various characterization techniques. A novel templating growth mechanism involving the use of Ir nanoparticles on the cathode surface may be responsible for the crystalline LiO_2 growth. Our results demonstrate that the LiO_2 formed in the Li-O_2 battery is stable enough that it can be repeatedly charged and discharged with a very low charge potential (~3.2V). We anticipate that this discovery will lead to future research into methods to synthesize and stabilize LiO_2. This can open the avenue for high-energy density lithium superoxide-based batteries as well as other potential uses such as a source of oxygen storage.

629861 Analysis of Shear Stress Profiles in a Novel Vertical Wheel Bioreactor for Human Stem Cell-Based Therapy

Abstract:

Human stem cell-based therapies represent one of the great promises of regenerative medicine to treat diseases and disabilities without a known cure. The Vertical-Wheel bioreactor describes a novel approach to expand and specialize stem cells homogeneously. A central aspect is its ability to provide a homogeneous hydrodynamic environment, with a uniform distribution of fluid forces, and lower shear stress levels, while also providing completely expanded stem cells with an optimal growth performance, maintenance of the high quality, karyotype and pluripotency of the initial cells. In this project, a Vertical-Wheel bioreactor simulation was created using a stationary laminar flow model governed by the Navier Stokes equations in COMSOL Multiphysics. The simulation provided shear rate and velocity profiles that were used to assess the homogeneity of the hydrodynamic environment inside of the bioreactor, along with the shear stress levels in the wall of the vertical-wheel impeller. The results indicate that by changing the inlet velocity, working volume, and impeller diameter, the shear stress levels at the wall of the impeller and the fluid velocity inside of the bioreactor changed, describing a different hydrodynamic environment. Thus, it is possible to predict the conditions of the hydrodynamic environment provided by the bioreactor with the simulation constructed, while also illustrating the different tendencies of shear rate and velocity profiles on computational fluid dynamics.

Modernizing the U.S. Electric Grid

A Proposal to Update Transmission Infrastructure for the Future of Electricity

Sara E. Murphy
American Institute of Chemical Engineers (AICHE)
11/2/21

WISE | WASHINGTON INTERNSHIPS *for* STUDENTS *of* ENGINEERING

University *of* Idaho

AIChE
The Global Home of Chemical Engineers

Table of Contents

Foreword

About the Author

Sara Murphy is a senior at the University of Idaho in Moscow, ID, majoring in chemical engineering and working towards a minor in political science. She has worked on several research projects during her undergraduate career with one involving bio-jet fuel and the other working to improve the performance of lead-acid batteries. In addition to her research, Sara has also worked as an energy engineer intern for the Department of Energy's Industrial Assessment Center program where she performs energy audits of local industries. Sara is the President of the Vandal Clean Energy Club, a member of the American Institute of Chemical Engineers, and a member of the Society of Women Engineers. Upon graduating in May 2022, Sara plans on pursuing graduate degrees in both engineering and public policy.

About the WISE Program

The Washington Internships for Students of Engineering (WISE) program is the collaborative efforts of several professional engineering societies to ensure that the future leaders of the engineering profession are aware of the important intersection between science, technology, and public policy. During their 9-week residency in Washington D.C., WISE interns meet with policymakers, industry leaders, civil servants, and representatives of nongovernmental organizations where they learn about the intricacies of the policymaking process. WISE interns are also tasked with choosing a specific scientific or technological issue and writing a public policy brief including policy recommendations related to the issue. The WISE program culminates with the publication of interns' policy briefs in the *Journal of Engineering & Public Policy* and their formal presentations on Capitol Hill. Find out more at https://wise-intern.org/.

Acknowledgements

I would like to thank the American Institute of Chemical Engineers (AICHE) for sponsoring my participation in the WISE program and their continued support throughout the program. Dr. Martin Abraham, my AICHE mentor also deserves acknowledgement for his supportive role during the development of my paper. I would also like to thank Mark Ames who served as the Faculty Member-in-Residence (FMR) this year and took the time to share his extensive knowledge of the public policymaking process with me and my fellow WISE interns. It was an honor to work alongside my fellow WISE interns and hear about their respective passions and insights. This public policy paper also would not have been possible without the encouragement of my family members, friends and my significant other, Andrew Bonilla. Thank you all for all your support, this would not have been possible without all of you.

List of Abbreviations

AC – Alternating Current

DC – Direct Current

DERA – Diesel Emissions Reduction Act

DOE – Department of Energy

DSO – Distribution System Operator

EIA – U.S. Energy Information Administration

ERCOT – Electric Reliability Council of Texas

EV – Electric Vehicle

FERC – Federal Energy Regulatory Commission

GW – Gigawatt

HVDC – High Voltage Direct Current

IEA – International Energy Agency

IED – Intelligent Electronic Devices

INL – Idaho National Laboratory

IOU – Investor-owned Utility

ISO – Independent System Operator

kV – Kilovolt

LLNL – Lawrence Livermore National Laboratory

LPO – Loan Programs Office

MW – Megawatt

NERC – North American Electric Reliability Corporation

NOAA – National Oceanic and Atmospheric Administration

NREL – National Renewable Energy Laboratory

OATT – Open-Access Transmission Tariff

RTO – Regional Transmission Operator

SERC – Southeastern Electric Reliability Council

SPP – Southwest Power Pool

TIFIA – Transport Infrastructure Finance & Innovation Act

TIP – Transmission Infrastructure Program

VRE – Variable Renewable Energy

WAPA – Western Area Power Administration

Abstract

Electric transmission infrastructure is often undervalued in the United States. However, a recent surge of electrical blackouts due to extreme weather conditions and natural disasters have brought the United States aging transmission infrastructure to the spotlight. Recent cyberattacks to critical energy infrastructure have also brought attention to the vulnerabilities in the U.S. electric infrastructure. Advancements such as growth of electrification in the transportation sector and the integration of variable renewable energy further, critical to decarbonization of the energy sector, further challenge the capacity of current electrical infrastructure.

Addressing deficiencies within U.S. electric infrastructure requires the collaboration of many stakeholders including federal agencies, state governments, local communities, and electricity generation and distribution entities. To properly undertake the challenges to the United States transmission infrastructure, it is recommended that potential solutions are accurately modeled, that the FERC uses its rulemaking capacity to foster collaboration and that the federal government creates national transmission loan program to specifically fund nationally significant transmission projects.

Why Does the Electric Grid Matter?

Electricity is an underemphasized asset within the United States. From heating and cooling our homes, to ensuring that all our smart devices are charged and ready for use, electricity is key. Within the past two years, access to reliable electricity has also empowered workers and students to continue their work and school obligations from home while COVID-19 safety protocols limited traditional attendance. Access to reliable electricity also enables the U.S. economy to produce goods and services and compete on the international market. The electric grid serves the purpose of transporting electricity from production to the American consumer, whether that consumer is a household or a production facility. In short, the United States electrical infrastructure is critical for powering all Americans' lives.

The Basics of the Electric Grid

The electric grid is a complex network that transmits electricity between power plants and the consumer. The components of the electric grid include power plants, transformers, transmission lines, substation transformers, distribution lines and load as highlighted in Figure 1 [2]. Power plants generate electricity by converting a fuel source like natural gas, sunlight, or wind to electricity. After power plants, the next stage is transformers where generated electricity is then converted from low voltage electricity to high voltage. Transmission lines then distribute this high voltage electricity within the grid and serve as an "interstate" system for electricity [2]. Electricity is carried by transmission lines to substation transformers which then "steps down" the voltage of electricity for distribution [2]. Distribution lines serve the load of an area and serve as the local system for electricity [2]. Load can be considered as the "consumer" of electricity whether the electricity be for lighting, powering up appliances, or heating or cooling our homes and businesses [2]. At this stage, the electrical consumption of consumers is metered and billed.

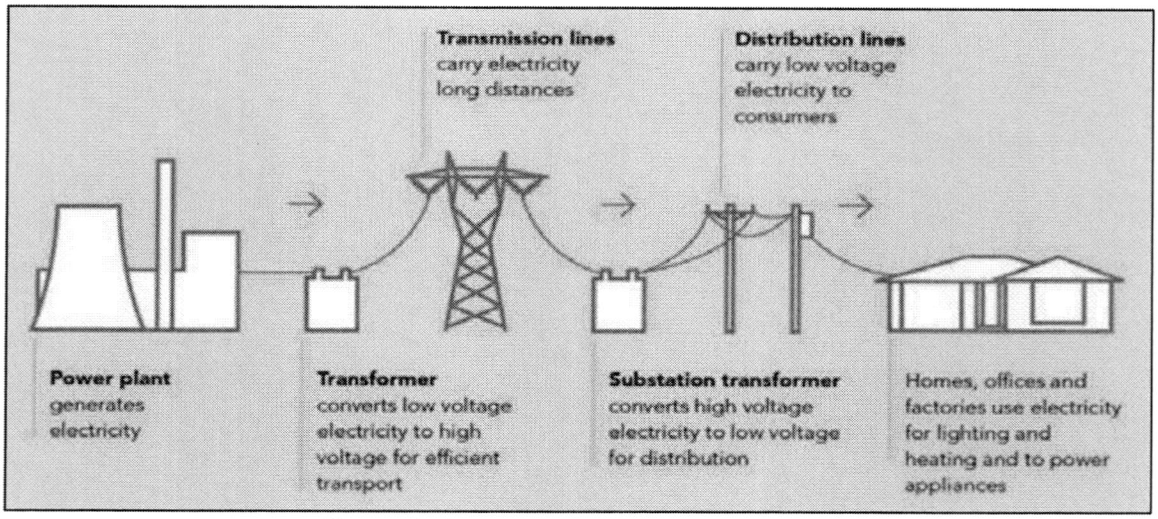

Figure 1: Components of the Electric Grid [2]

For the scope of this policy paper, the focus is going to be on updating transmission infrastructure of the electric grid. This focus on transmission is due to its importance in transporting electricity over long distances, from generator to consumer and from rural areas with generation capacity to urban areas with high load. However, before describing transmission infrastructure, the concepts of current and voltage must be defined. In this context, current is defined as flow rate of electric charge and voltage can be thought of as the "electrical pressure" within a wire [2]. The transmission system consists of alternating current (AC), direct current (DC) and high voltage direct current (HVDC) lines. DC lines work by supplying a constant voltage while AC lines alternate voltage over time. AC lines have less power loss, which makes them more efficient and often preferred over DC lines for spanning long distances. While DC and AC transmission lines are more prevalent, the use of HVDC is growing. The prevalence of HVDC lines is expected to continue growing as it is more efficient to transport high voltage electricity over long distances. HVDC systems also contain technology that can convert AC and DC voltage and can be used to support existing systems and new infrastructure. It is for these reasons that HVDC lines and systems are seen as a possible solution to the current challenges that plague U.S. transmission infrastructure. Focusing solely on updating transmission infrastructure allows the U.S. to take one step towards creating a more robust, interconnected grid, without being overwhelmed with other challenges.

The Human Element

Limited access to electricity can affect people's livelihoods and even their health. This is especially true when electrical blackouts occur in extreme weather events such as heat waves or winter storms. These blackouts can occur when there is damage to electrical infrastructure or equipment failure at a power plant. Electrical blackouts may also occur when the load overwhelms what the utilities can supply, resulting in complete blackouts or "rolling blackouts" where electricity is deliberately turned off in several areas sequentially for a specified amount of time to reduce demand. Brownouts are like "rolling blackouts" as they also reduce electrical load but do so by limiting the level of voltage to homes and businesses. Load tends to become an issue in extreme weather events since individuals increase their use of air conditioning or heating to remain comfortable and therefore strain the grid.

A research study from the journal *Environmental Science & Technology* found that blackouts have increased by 60% since 2015 [3]. This trend is illustrated in Figure 2 [4] which shows the amount of major blackout events and the average duration of said events between the years of 2015 to 2019. Blackouts in conjunction with heat waves are of particular concern, since air conditioning is not an option, and the extreme heat can cause health problems like heat exhaustion or heat stroke. These researchers estimated that two-thirds of residents from Atlanta, Georgia, Detroit, Michigan, and Phoenix, Arizona would be at risk for heat-related health conditions if a heat wave and a blackout were to occur simultaneously [3], [4].

Additionally, it was noted that cooling centers which are powered by backup generators, could only provide for 2% of these three cities residents, leaving many residents at risk and in the dark [3], [4].

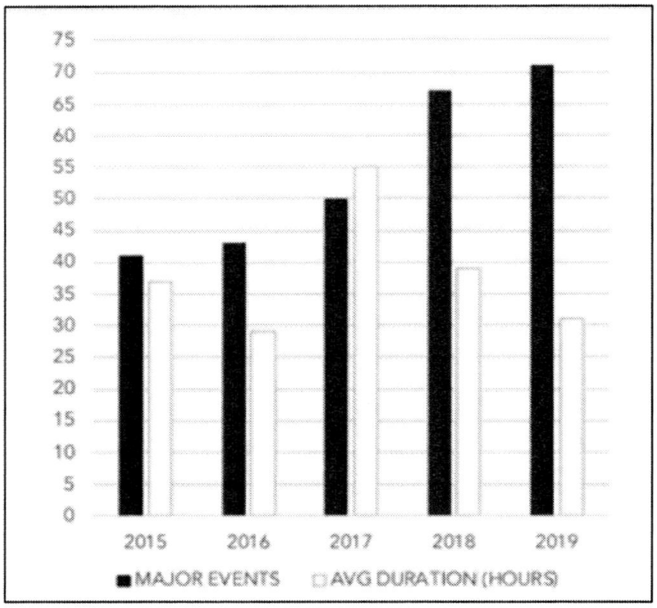

Figure 2: Total Number of Major Electrical Grid Failure Events and Average Event Duration (Hours) for U.S. Power Utilities (2015-2019) [5]

Utilizing Texas as an additional case study, it is evident that no matter the type of extreme weather event, the electric grid needs to be updated to be more robust and better connected. For example, in February 2021, 69% of Texans lost electricity for an average of 42 hours while the temperature in some areas reached below freezing [5]. Part of the reason these blackouts occurred in Texas was because natural gas pipelines froze and diesel engines would not start, thus forcing power plants to go offline in the extreme winter conditions [6]. Texas also faced similar problems with heat waves in the summer months of 2021. For instance, in late June 2021, Texas residents were asked to place their thermostats no lower than 78°F in an effort to reduce the electrical demand on the grid [7]. The use of Texas as a case study is key because it highlights how a state's isolation from other generation sources can be catastrophic in extreme weather conditions. By modernizing and interconnecting current transmission infrastructure, states like Texas would be supported by additional electricity supply and the number of Texans and Americans that experience blackouts every year could be reduced.

Why Update the Grid Now?

Growth of Electrification

Figure 3 [8] highlights the historical electricity consumption per sector in the United States from 1950 to 2020 and 2021-2050 projections of anticipated electricity consumption. One notable trend is the relatively limited growth of electricity consumption in the commercial, residential and industrial sectors [8]. This limited growth can be attributed to the recent focus on energy efficiency in these sectors. Utility companies have further galvanized energy efficient measures by offering rebates and incentives to facilities that are energy efficient. Many of these measures also benefit businesses as the reduction in energy use also leads to cost savings. Limited growth in the industrial sector can also be credited to the recent shift offshore of American manufacturing.

An important trend in Figure 3 is the projected increase of electricity consumption in the transportation sector. Said growth is expected to continue as electric vehicles (EVs) become more prominent in personal use and in public transportation. For example, in 2020 1.8 million EVs were registered, which was three times the amount that were registered in 2016 [9]. Support for EVs is partially due to recent pushes to reduce greenhouse gas emissions in the transportation sector. President Biden emphasized this as well when in August 2021 he signed an executive order that aims to make half of the American car sales electric cars by 2030 [10]. Additionally, a study conducted by Lawrence Berkley National Laboratory predicts that battery

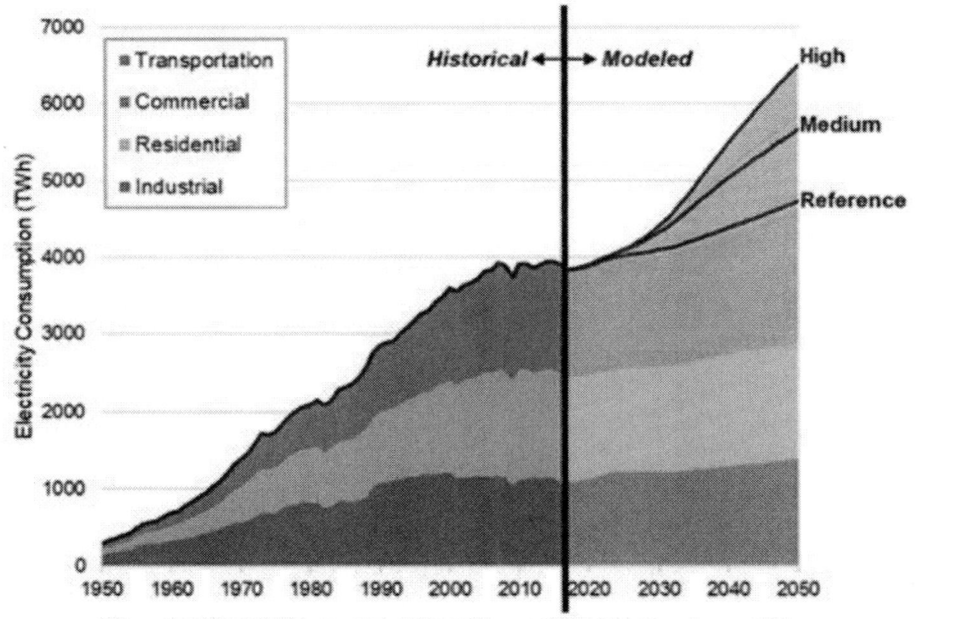

Figure ES-3. Historical and projected annual electricity consumption

Moderate technology advancements are shown. Slight adjustments were made to the modeled industry consumption estimates (for 2017–2020) to align them with available historical data.

Figure 3: United States Historical and Projected Annual Electricity Consumption [8]

costs will continue to decline and thus EV prices will also continue to decline, allowing EVs to be a more feasible purchase for the average American [11]. Car manufacturers are also getting on board as General Motors recently announced it will no longer be selling gasoline-powered passenger vehicles by 2035 and will be expanding its offerings of electric vehicles [12]. There are also long-term advantages for individuals who consider purchasing an EV over a traditional gasoline vehicle. One study conducted by the DOE found that an EV is about 4 cents per mile less expensive than the traditional gas vehicle when considering common maintenance costs [13]. While at first this does not sound like substantial savings, the savings add up as more miles are driven. For example, during the average lifetime of an internal combustion engine, about 200,000 miles, it will cost about $20,200 to maintain a traditional gasoline vehicle while it would cost $12,200 to maintain an EV [13].

Electrification of public transportation is also expanding more than previously expected. This is especially the case after the Biden Administration's call for 20% of the nation's school buses to be electric [12]. Other policies such as the Low-or-No Emission Grant program or the Diesel Emissions Reduction Act (DERA) grants have further mobilized the use of EVs for public transportation. Many municipalities have also started utilizing electric buses due to their long-term cost saving opportunities despite the fact that the initial cost is double that of a traditional, diesel bus [12]. These long-term cost savings allow public transportation departments to evaluate the benefit of building an electric fleet now, for commuters of the future.

This is not to say that electric transportation doesn't have room for technical improvement. For EVs to truly dominate the market, on-board storage capacity will have to continue to increase to extend the current range. There will also have to be faster recharging rates for long distance applications. Updates to the current electric grid will be critical to increasing the role of EVs in transportation. As researchers from Lawrence Livermore National Laboratory (LLNL) found, modern power infrastructure is vital for EVs to make up a large fraction of the light-duty vehicle fleet [11]. Namely, updates to transmission infrastructure would assist in establishing widespread EV charging infrastructure. Overall, by modernizing the electric grid, the United States has the opportunity of ushering in a new era of electric transportation with reduced greenhouse gas emissions. However, without updating the current electric grid, this future will be difficult to obtain.

Changing Electricity Mix

In the last 10 years, the United States has drastically shifted what sources it uses to produce electricity. As seen in Figure 4 [14], natural gas surpassed coal-fired power plants as the predominant source of electricity production in 2015. The U.S. Energy Information Administration (EIA) reported that over 121 coal-fired power plants were retired or repurposed between 2011-2019, with 103 being converted in natural gas-fired plants [15]. The EIA attributes the shift from coal to natural gas power generation to three factors: (1) stricter emission standards, (2) low natural gas prices and (3) updated gas turbine technology [15]. In contrast to the changes in the use of coal and natural gas for electricity production, nuclear and hydropower sources have remained about constant over the last 10 years.

Another notable trend includes the expansion of renewable energy sources in producing electricity. Like natural gas, the growth of solar and wind's generation can be attributed to recent reductions in cost for the necessary materials and the demand for low and no emission electricity sources. While integrating wind and solar energy into the electricity grid is not a significant obstacle right now, it is predicted to be one in the future. EIA projects that by 2050 wind and solar energy could account for as much as 33% of the total electricity generation [16], which is expected to create challenges.

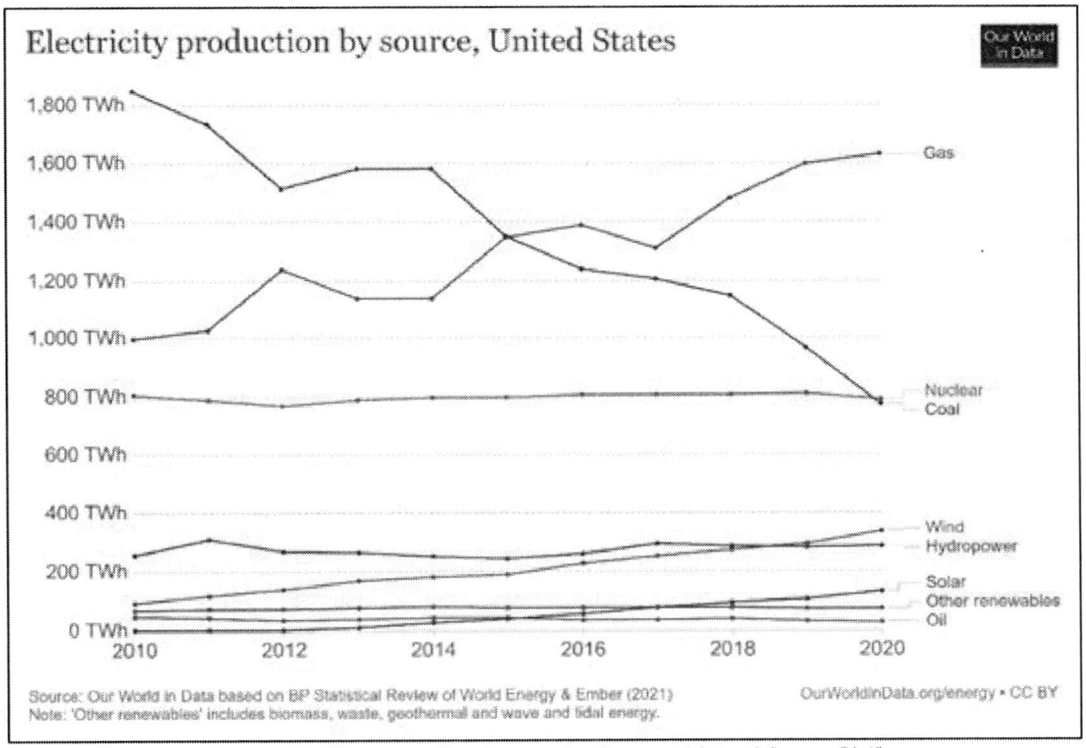

Figure 3: Electricity Production by Source, United States [14]

Wind and solar energy are termed as variable renewable energy (VRE), which implies that source can only produce electricity when the conditions are right. One of the challenges with VREs is that they are frequently built in rural regions of the United States, as these areas have the best conditions to produce significant amounts of electricity. Unfortunately, the transmission infrastructure is often lacking in these areas, local communities are unable to consume all the electricity produced, and there is minimal storage capacity. Updates to transmission infrastructure would allow electricity generated in rural communities to be transmitted over longer distances and utilized in high demand, urban areas.

VREs are also difficult to merge into the electricity mix because the electric grid was designed to handle "on demand" forms of electricity, and not the intermittent nature of wind and solar generation [16]. For instance, wind speeds are frequently greater at night and thus wind farms produce more electricity, despite electricity consumption being lowest at night. And current U.S. electrical infrastructure lacks the storage capacity to store this electricity for use at another time. In short, VREs face the problem of matching up high generation times to high consumption times. Improving the transmission infrastructure would allow generated electricity to quickly reach consumers throughout the U.S. with less concern about overlapping VRE's high generation with high consumption times.

Another challenge that makes integrating VREs into the electricity mix difficult is grid congestion. Grid congestion occurs when existing transmission or distribution lines are unable to accommodate high demand or high generation conditions. The best metaphor for this is a traffic jam of electrons on the transmission line. Grid congestion often limits the amount of renewable energy projects that are viable within an area. One example of this occurred in the Midwest region of the U.S., where 245 clean energy projects were halted between 2016-2020 at advanced development, with the majority of the reasons being the lack of transmission capacity [17]. From this example it can be extrapolated that the core challenge for integrating VREs is not at the generation stage, but the transmission phase.

One idea to modernize the grid is to build a national HVDC transmission system. NOAA scientists found that such a grid would allow 523 GW of new wind and 371 GW of new solar to be incorporated [18]. To put these values into perspective, 1 GW is about 412 utility-scale wind turbines and can power about 110 million LED lightbulbs [19]. An alternative solution would be to increase storage capacity to hold excess VRE from the time it is generated to the time it is needed. Unfortunately, technology advancements have not yet made electric storage a viable solution and thus the current focus should be on improving U.S. transmission infrastructure.

The Regulatory Environment of Electricity

Regional Electric Grids

The contiguous United States transmission system consists of three regional power grids, or interconnections, as seen in Figure 5 [20]. Bulk power transmission between and reliability of the interconnections is regulated by a non-profit called the North American Electric Reliability Corporation (NERC). NERC includes parts of both Canada and the United States, making the issue of updating transmission infrastructure a national and international challenge. The two major AC grids of NERC include the Western Interconnection and the Eastern Interconnection while the minor one is the ERCOT Interconnection. ERCOT stands for the Electric Reliability Council of Texas and was created by Texas utility companies to avoid federal regulations.

Because the electric grid in the United States is broken up into regional grids there is limited exchange of electricity between the grids. In turn this means that the three U.S. interconnections operate in isolation. This isolation is problematic when the electric generation within each individual grid struggles to meet the demand within the region. Understanding the organization of electricity regulation is critical to be able to advocate for future policies affecting transmission planning and electrical infrastructure.

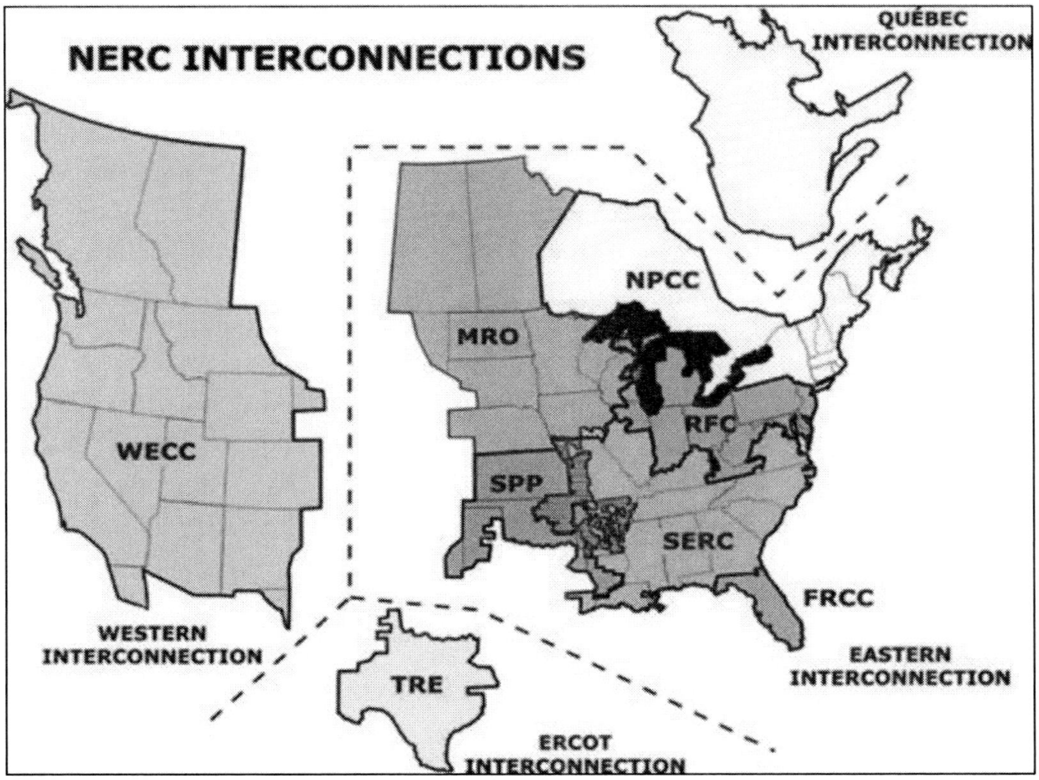

Figure 5: North American Electric Reliability Corporation (NERC) Interconnections Map [20]

Prior to the 1990s, the majority of U.S. residents received their electricity from investor-owned, vertically-integrated monopolies otherwise known as IOUs [21]. During this time IOUs handled electricity generation, transmission planning, local distribution, and billing. However, reforms in the 1990s and 2000s limited the power of IOUs by separating generation owners from the transmission owners and both from distribution system operators (DSO) [21]. Electricity consumers saw these reforms as an opportunity to gain access to the wholesale market while utility companies faced the reality that lower consumer prices resulted in "equivalent reductions in returns for utility shareholders" [21]. As a result of these reforms, transmission planning became the responsibility of independent system operators (ISOs) and regional transmission operators (RTOs). Figure 5 highlights the various ISOs and RTOs within the United States**Error! Reference source not found.**. The black cross-hatched region in Figure 5 in between the Southwest Power Pool (SPP) and the Southeastern Electric Reliability Council (SERC) represents an overlap of jurisdiction between the two regional entities. DSOs also known as utility companies are not included in Figure 5 as they depend on the locality and can differ even within a state. While ISOs and RTOs handle large scale transmission planning, DSOs manage metering and billing for their respective customers.

The Role of Federal Government

The Department of Energy (DOE) is the main department within the U.S. government that oversees energy including electricity and electric infrastructure. Within the DOE, there are many programs, agencies, and laboratories. DOE's national laboratories such as the National Renewable Energy Laboratory (NREL) in Golden, Colorado serve as support to the federal government and state governments by providing research studies and specialty experts. Published reports by the national laboratories also assist in informing the public and policymakers about electricity trends as well as in modeling potential solutions to both current and future transmission challenges.

Regulation of energy is handled by the Federal Energy Regulatory Commission (FERC) which is a commission within the DOE. FERC was created to replace the Federal Power Commission that was formed to oversee hydropower dams in the 1920s. In 1935, Congress passed the Federal Power Act which dismantled utility monopolies and required the Federal Power Commission to set "just and reasonable" wholesale electricity prices [22]. These duties were then carried over to the mandate of FERC which was created in 1977. FERC has regulatory authority over the interstate sale of electricity and operation of regional markets [22]. FERC does this by approving the terms and rates of interstate transmission service as described in the transmission owners open access transmission tariff (OATT) [23]. OATT requires all public utilities that own, control, or operate facilities used for transmitting electricity in interstate commerce to file open access non-discriminatory transmission tariffs [24]. FERC's goal with

OATT was to remove impediments to competition in the wholesale bulk power market and bring more efficient and lower cost power to the U.S. electricity consumers [24].

Congress has the ultimate authority over FERC and can call FERC commissioners, or senior staff to testify before various congressional committees [25]. NERC, as previously discussed, operates beneath FERC and oversees the more technical aspects such as power system reliability, operation as well as transmission planning [22]. Due to NERC's more hands-on involvement with electricity it often works with ISOs, RTOs, and DSOs alike.

Overall, FERC mostly serves as a rulemaking commission and is not involved in the technical aspects of electricity. FERC Order 890 outlined the general requirements for local and regional transmission planning practices and procedures. Order 1000 served as an addition to the previous Order 890 while also correcting what the previous order lacked, such as requiring public utility transmission providers to participate in a regional transmission planning process [26]. This involved instructing these providers to work together with their respective partners in their transmission planning region to determine more efficient or cost-effective transmission plan to meet both their needs [26]. Regional transmission planning efforts were also expected to have a regional cost allocation method for new transmission facilities that met all six regional cost allocation principles [26]. Similar reforms to this were also made for interregional transmission planning efforts. The most contentious reform in Order 1000 was the one that allowed non-incumbent transmission developers to complete transmission projects within the service area of incumbent utilities [23]. This reform was created in hopes it would drive competition within these areas and that non-incumbent developers could build transmission infrastructure at lower costs. However, this reform was not well-received by incumbent utilities. While FERC's Order 890 and Order 1000 did increase spending on transmission, there are still concerns with the United States' electric transmission infrastructure. These concerns include: the lack of transmission projects that independent developers can bid on and that the current investment into transmission is not enough to meet future electricity needs [23]. Such concerns among others are why electric transmission infrastructure remains a consequential issue and why new policies must be written, and new technology must be implemented.

Key Conflicts and Concerns

Conflict of Stakeholders' Interests

There are many stakeholders involved when it comes to the issue of modernizing the electric grid. Governmental stakeholders include the U.S. federal government, state governments, and local governments, whose competing interests make planning national electric infrastructure challenging. This is especially the case as interregional and national transmission planning becomes more incentivized by FERC. At the federal level, FERC oversees the inter-state sale and price of electricity and is also a rulemaking body that sets requirements for regional and interregional transmission planning and development. NERC also operates at the federal level, beneath the guise of FERC. At the state level, ISOs and RTOs regulate transmission planning and the wholesale energy markets within the restructured areas of the U.S. grid. Additionally, state governments control the consumer prices within each respective state. Each state also has its own regulatory environment that can either support privatizing the electric grid or regulating it in efforts to protect the consumer. These various and conflicting regulatory environments is what makes it challenging to establish an interregional or national transmission plan within the United States. Local governments also influence the electric grid since transmission projects often involve various state and local agencies that determine transmission siting in states and localities [27]. Like state governments, local governments also have distinct regulatory environments that dictate their policies towards the electric grid, which is similarly problematic when integrating various localities' grids.

Besides governmental stakeholders, other entities such as utility companies or DSOs, landowners and constituents take part to determine if certain transmission projects occur in their local communities. For instance, transmission developers often struggle to find landowners who welcome the prospect of having new transmission lines built nearby [27]. Additional complexity is added when planned transmission routes pass through federal or tribal lands [27]. When it comes to regional utility companies, they have vested interest in both maintaining the current electric grid and in modernizing it. On one hand, maintaining the current electric infrastructure benefits utility companies by limiting the amount of large capital investments needed to modernize the grid. Also, modernizing and further integrating the electric grid, may threaten some utility companies that are concerned with competing on the national market. On the other hand, current transmission infrastructure limits generation capacity and the ability to get electricity where it needs to go. It is for this reason that some utility companies view investing in modern transmission infrastructure as a profitable venture. However, utility companies still have concerns including the ability to generate profit for investors, justifying large capital investments, and wanting to minimizing risk. These concerns as well as state and local government interests can be addressed by incentivizing utility companies and local and state governmental agencies to partake in interregional and national transmission planning.

Security Concerns

A more connected grid will require a higher level of security from both physical attacks on infrastructure and cyberattacks on computer systems. Cyberattacks are of particular concern as these attacks have become more prevalent as computers and embedded devices have become more integrated into critical electrical infrastructure. A report by IBM in February 2021 found that the energy industry was the third most targeted sector in 2020 by cyberattacks [28]. The heightening of electricity-related cyber incidents worldwide is shown in Figure 6 in the light blue color [29]. As stated previously, cyber vulnerabilities of the grid are largely due to the integration of intelligent electronic devices (IEDs), however other vulnerabilities can exist because of outdated equipment or software [30]. The portions of the electric grid that are expected to be the most targeted by cyberattacks include generation and transmission infrastructure [30]. This does not mean that distribution is not vulnerable to cyberattacks, only that an attack on the other components would produce more detrimental and cascading consequences.

Preventing and protecting the electric grid from cyberattacks has become a recent priority of the Biden administration. In April 2021, the administration released a 100-day plan outlining how power utilities will be incentivized to install new monitoring equipment that will be able to quickly detect hackers and alert the U.S. government [31]. Other suggested actions by the International Energy Agency (IEA) involve developing policies that foster sector-wide collaboration and response procedures and setting up research partnerships with industry and academia to promote R&D in cyber resilience of the U.S. electric grid [29]. From this discussion it is evident that any future efforts to modernize the electric grid must not only focus on physical infrastructure but also on the security of the grid.

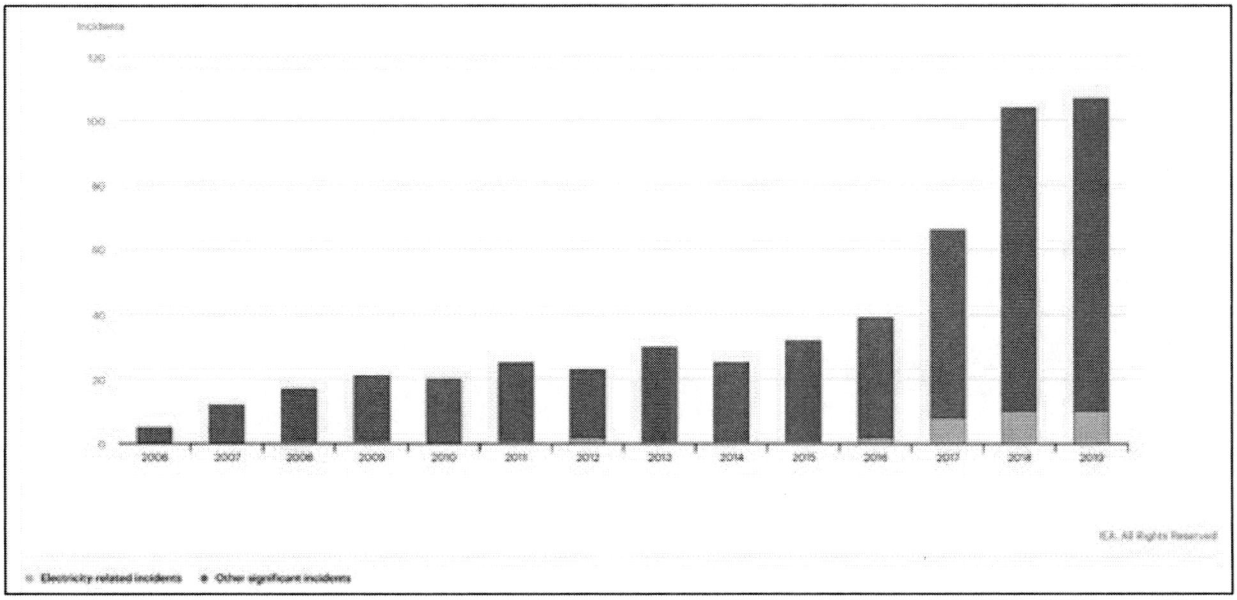

Figure 6: Significant cyber incidents worldwide, 2006-2019 [29]

Policy Recommendations

Model Potential Solutions

One policy recommendation is to continue funding research programs that study and model the electric grid and assist in determining the most cost-effective solutions. Various models that highlight their differing benefits and detriments allow stakeholders to collaborate and establish a solution that is satisfactory to all parties. For example, the *Interconnections Seam Study* conducted by the National Renewable Energy Laboratory (NREL) features a series of potential ideas to modernize the United States' current transmission infrastructure [32]. The study analyzed four separate scenarios and each of these scenarios' respective transmission lines are shown in Figure 7 [20].

Design 1, for instance, includes no new HVDC lines or capacity but adds more AC lines, which are represented in Figure 7 as short black lines [20] represents the standard design to which the other designs included in the study were compared [20], [32]. The purpose of Design 1 is to increase transmission of electricity within each respective interconnection. Design 2a, on the other hand, increases the capacity of the HVDC ties and adds additional AC lines [20]. Increased HVDC capacity is represented as red 'butterflies' along the dotted line between the interconnections in Figure 7. While Design 2a doesn't include the installation of new HVDC lines, it does increase the capability of electricity to be exchanged between the regional interconnections. In contrast to Design 2a, Design 2b adds HVDC lines with a few AC lines

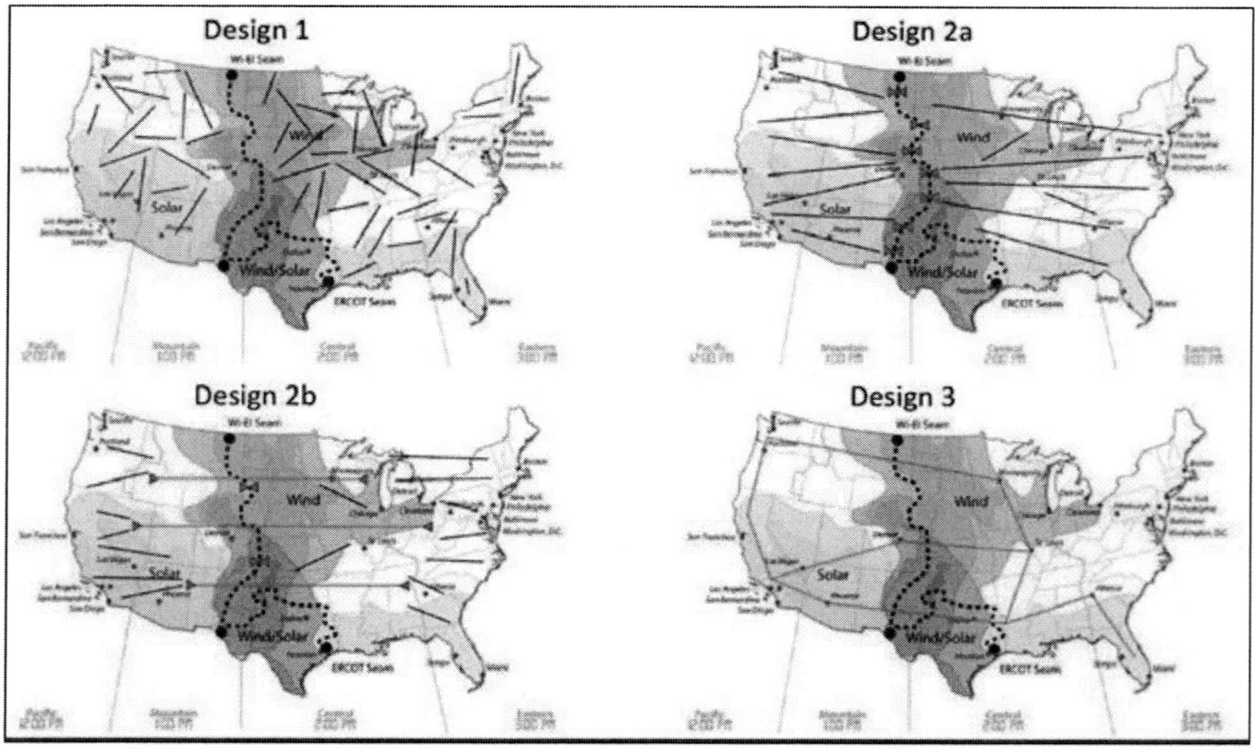

Figure 7: Transmission Designs in the Capacity Expansion Work [20]

branching off [20], [32]. These additional HVDC lines are portrayed in Figure 6 as red lines. While Design 2b establishes three HVDC lines that stretch across the U.S., Design 3, models a complete, national HVDC grid with some new AC lines [20], [32]. Nonetheless, all of the designs proposed in the NREL study were found to reduce electricity costs and meet reliability and demand requirements [20].

The *Interconnections Seam Study* also incorporated a cost-benefit analysis of each of the designs in the study. Said analysis utilized the investment costs and operating costs for the years between 2024 – 2038 and an additional 20 years with no increase in load or generation after 2038 [33]. In addition, since Design 1 had limited adjustments to our current transmission infrastructure it was used as the standard design to compare Designs 2a, 2b and 3 [33]. This is represented in Table 1 as the delta notation before designs 2a, 2b and 3. As stated previously, the cost-benefit analysis included examination of investment costs, generation investment costs and operation cost with each being reported in billions of dollars. Additionally, the analysis determined the 35-year net cost charge which is the difference between the gross cost of operation and investment and the potential benefits gained from the cost. Also shown in Table 1 is the benefit cost ratio over a 35-year period to express which designs have more respective benefits when compared to Design 1.

Table 1: Summary of CGT-Plan Benefit/Cost Results for Base Scenario [33]

Capacity or Cost Item	D1	ΔD2a	ΔD2b	ΔD3
Transmission Investment Cost, $B	40.03	2.57	6.76	8.19
Generation Investment Cost, $B	555.23	3.6	10.44	4.17
Operation Cost, $B	2,376.50	-8.79	-21.70	-15.30
35-yr Net Cost Charge, $B	-	-2.62	-4.5	-2.94
35-yr B/C Ratio	-	2.02	1.66	1.36

One thought-provoking finding from the study is that Designs 2a, 2b and 3 all provide cost savings relative to Design 1 and illustrate the financial advantages of investing in each of the modeled designs. From this analysis it also appears that Design 2a has the greatest benefit to cost ratio within 35 years therefore emphasizing the need to increase HVDC capacity along the seam between the interconnections. Overall, studies like the *Interconnections Seam Study* are useful for clearly demonstrating the costs and benefits of various potential designs for modernizing the United States' transmission infrastructure. This study and similar studies can also be employed to highlight the advantages of transmission updates to stakeholders who are hesitant to invest as well as assist in sparking public support for updated transmission infrastructure. Continued or increased support to further explore opportunities to improve transmission infrastructure will benefit decision-makers in deciding among competing alternatives.

Require Cooperation in Transmission Planning

Besides modeling potential transmission solutions, it is also critical for more cooperation to be established between the stakeholders involved in transmission planning. Cooperation, however, is challenging to create since reforms in the 1990s and 2000s consolidated oversight of transmission planning to IOUs. As a result, the transmission planning process has remained nontransparent and unchecked, with limited access for journalists and public interest groups to investigate and report on the transmission planning process.

FERC has attempted to promote more cooperation in transmission planning through its rulemaking capacity. For instance, in 2007 FERC Order 890 set the baseline for regional planning and open-access transmission service [23]. However, by 2010, FERC recognized that Order 890 was lacking and utilized Order 1000 (2011) to further strengthen the planning process [23]. Order 1000 both expanded on requirements for public utility transmission providers to participate in regional transmission planning and for them to satisfy the nine principles contained in the prior Order 890 [23] [26]. These nine principles include: coordination; openness; transparency; information exchange; comparability; dispute resolution; regional coordination; economic planning studies; and cost allocation [23]. FERC's main goal with utilizing its rulemaking capacity in this way was to foster collaboration between various stakeholders to thus develop cost-efficient transmission developments that would benefit all involved parties and distinct regional portions of the electric grid.

Despite previous rulemaking by FERC, IOUs still maintain control over transmission planning. This has allowed IOUs to focus on local developments rather than regional planning. FERC should require IOUs to be a part of RTOs to foster greater regional cooperation [34]. FERC making this a requirement is significant as IOUs currently have substantial leverage over RTOs by being able to pull out of their respective RTO at any time. By requiring IOUs to be a part of RTOs, RTOs would have more capability to act as a regional body, foster collaboration between various IOUs, and ensure more regional transmission planning occurs.

FERC should also have greater oversight of IOUs transmission planning efforts by holding IOUs responsible for justifying the costs of transmission projects [34]. Instead of shifting the burden onto IOUs, FERC could also involve or create an independent agency that would assess the proposed costs for transmission projects. Said FERC rulemaking would establish a precedent of more cost-effective spending on and expect IOUs to be both more transparent and collaborative with their development of transmission projects.

Finance Significant Transmission Projects

The Loan Programs Office (LPO) within the Department of Energy serves as a financer for large scale energy projects throughout the United States. These projects and the respective companies funded through LPO have led to the creation of jobs and the enabling of budding industries to grow. One example, Tesla Inc. received funding through the LPO in the form of loan guarantees which were then utilized to develop their facility in Fremont, California [35]. This facility then went on to produce the Model S sedan and usher Tesla in as a world leader in electric vehicles [35].

When it comes to financing transmission projects, the Title 17 Innovative Energy Loan Guarantee Program and Tribal Energy Loan Guarantee Program allows the LPO to distribute about $3 billion for projects that involve innovative technology such as HVDC lines, off shore wind transmission, and electrical infrastructure along transportation corridors determined by the Department of Transportation [36]. In addition, Title 17 also appropriates $2 billion in partial loan guarantees to projects that are owned by federally recognized tribes or the Alaska Native Corporation [36]. However, the LPO's method of financing of transmission projects falls short. The first problem is the amount of funding being allocated to transmission projects. $3 billion is simply not enough funding to complete larger scale, collaborative projects that are required to create a more robust, resilient grid. For example, transmission projects planned for the New England region had estimated price points of between $1.2 billion and $2.2 billion in 2013 [37]. While modern technology may have slightly reduced this price point, it still unrealistic to think that $3 billion in loan guarantees can spur the national transmission planning and development that the U.S. so desperately needs.

In April 2021, the Biden Administration began to address some of the challenges with financing transmission projects. This was done by announcing the availability of up to $8.25 billion in loans from the LPO and the Western Area Power Administration (WAPA) [38]. Unlike the LPO, WAPA works by awarding loan guarantees through its Transmission Infrastructure Program (TIP) which was created to incentivize "the delivery of reliable, affordable, clean power in the Western United States" [38]. The main goal of this additional funding was to encourage efforts to expand and update the nation's transmission infrastructure [38].

Federal policymakers have also begun to see the need to modernize current transmission infrastructure. H.R. 3684 otherwise known as the Infrastructure Investment and Jobs Act contains extensive policy related to U.S. transmission infrastructure [39]. As of October 2021, H.R. 3684 has been passed by both the U.S. House of Representatives and Senate, and now is in the phase of reconciling differences and amending prior to being sent to President Biden's desk [39]. The significance of H.R. 3684 is that it addresses a myriad of issues that involve the United States' outdated transmission infrastructure through the creation of federal programs and the allocation of grants. For instance, policy in H.R. 3684 directly confronts issues regarding electrical blackouts by defining what a disruptive event is, how they are frequently related to natural disasters, extreme weather, and wildfires [39]. H.R. 3648 also details the creation of a federal financial assistance program titled the "Upgrading Our Electric Grid and Ensuring Reliability and Resiliency" [39]. Said program would provide competitive financial assistance to and collaborate with electric sector owners and operators to "demonstrate innovative approaches to transmission" and other technologies [39]. Cybersecurity is also addressed in H.R. 3684 with the creation of several programs including "Energy Cyber Sense" program and technical assistance programs for rural and municipal utilities [39].

A good avenue for additional funding for transmission infrastructure would be to create a national transmission loan program modeled like the Transportation Infrastructure Finance and Innovation Act (TIFIA), which was created to prioritize transportation projects that are regionally or nationally significant [40]. Said program would incentivize interregional and national transmission planning as well as increase the likelihood that large-scale transmission projects would be funded and completed. It would also provide a wider variety of funding options which would provide more flexibility and thus stimulate a more competitive market and likely produce more cost-effective transmission projects. In general, a national transmission loan program would bring U.S. transmission infrastructure into focus and would serve as a mechanism to finance up-and-coming developments that will be crucial to updating current transmission infrastructure.

Conclusion

Access to on-demand electricity is both expected and needed throughout the United States whether it's by a homeowner, a production line, or a hospital. However, access to reliable electricity has been facing challenges related to the aging electrical infrastructure of the U.S., and its need to be reformed. Difficulties with dated transmission infrastructure have been further exacerbated by the progress in other sectors such as the growth of electrification in transportation and the role of VREs in electric generation. While these developments are essential to decarbonize the United States' energy sector, they have further highlighted the deficiencies of current infrastructure.

The relationship between electrical blackout events and extreme weather also emphasizes that the limited resiliency of current infrastructure is not only a technological issue, but also a potential health concern. Other concerns regarding the security, specifically the cybersecurity of critical infrastructure like electricity, have further established the need to modernize transmission infrastructure to be able to prevent attacks from bad actors.

Due to the vastness of electrical infrastructure in the United States, it is not a simple undertaking and will require the collaboration of many stakeholders at both the federal, state, and local level as well as entities from the private sector. To begin addressing the various deficiencies of present transmission infrastructure, it is recommended that:

(1) Potential transmission designs are modeled, and their respective costs and benefits evaluated

(2) FERC utilize its rulemaking capacity to cultivate collaboration among stakeholders, specifically between IOUs and RTOs by requiring all IOUs are members of their appropriate RTO

(3) Additional funding and attention be allocated to U.S. transmission infrastructure, through the creation of a national transmission loan program modeled the well-established TIFIA programs

Fortunately, policymakers have started to identify the importance of electric infrastructure within the United States, with both members of Congress and the Biden Administration creating policy that allocates funding to aging infrastructure. The work of the Biden Administration and success of H.R. 3684 reveals that policymakers are ready to tackle the nationally significant challenges such as electric infrastructure, which serves a hopeful focal point for the future of U.S. electricity.

References

[1] "2020 Standard Scenarios Outlook Models Possible Evolution of U.S. Electricity Sector with Expanded Metrics," *National Renewable Energy Laboratory*, Dec. 03, 2020. https://www.nrel.gov/news/program/2020/2020-standard-scenarios-outlook-models-possible-evolution-of-us-electricity-sector-with-expanded-metrics.html

[2] A. McCoy, "Transmission 101: The Fundamentals Of High Voltage Transmission," presented at the Electric Transmission Infrastructure 101, Feb. 16, 2017. Accessed: Jul. 19, 2021. [Online]. Available: https://www.eesi.org/briefings/view/021617wires

[3] C. Flavelle, "A New, Deadly Risk for Cities in Summer: Power Failures During Heat Waves," *The New York Times*, May 03, 2021. Accessed: Jun. 21, 2021. [Online]. Available: https://www.nytimes.com/2021/05/03/climate/heat-climate-health-risks.html

[4] B. Stone *et al.*, "Compound Climate and Infrastructure Events: How Electrical Grid Failure Alters Heat Wave Risk," *Environ. Sci. Technol.*, vol. 55, no. 10, pp. 6957–6964, Apr. 2021.

[5] K. Watson, R. Cross, and M. Jones, "The Winter Storm of 2021," Hobby School of Public Affairs, University of Houston, Mar. 2021. Accessed: Jun. 22, 2021. [Online]. Available: https://uh.edu/hobby/winter2021/storm.pdf

[6] W. Englund, "The Texas grid got crushed because its operators didn't see the need to prepare for cold weather," *The Washington Post*, Feb. 16, 2021.

[7] J. Schwartz and I. Penn, "Texas Power Grid, Strained Last Winter, Now Faces an Early Heat Wave," *The New York Times*, Jun. 15, 2021. Accessed: Jun. 22, 2021. [Online]. Available: https://www.nytimes.com/2021/06/15/climate/texas-heat-wave-electricity.html?searchResultPosition=1

[8] T. Mai *et al.*, "Electrification Futures Study: Scenarios of Electric Technology Adoption and Power Consumption for the United States," National Renewable Energy Lab (NREL), 2018. Accessed: Jun. 28, 2021. [Online]. Available: https://www.nrel.gov/docs/fy18osti/71500.pdf

[9] D. Desilver, "Today's electric vehicle market: Slow growth in U.S., faster in China, Europe," *Pew Research Center*, Jun. 07, 2021. https://www.pewresearch.org/fact-tank/2021/06/07/todays-electric-vehicle-market-slow-growth-in-u-s-faster-in-china-europe/ (accessed Jun. 19, 2021).

[10] D. Shepardson and J. Mason, "Biden seeks to make half of new U.S. auto fleet electric by 2030," *Reuters*, Aug. 05, 2021. Accessed: Aug. 19, 2021. [Online]. Available: https://www.reuters.com/business/autos-transportation/biden-set-target-50-evs-by-2030-industry-backs-goal-2021-08-05/

[11] P. Jones, J. Levy, J. Bosco, J. Howat, J. Van Alst, and L. Schwartz, "The Future of Transportation Electrification: Utility, Industry and Consumer Perspectives," Lawrence Berkley National Laboratory, Report No. 10, Aug. 2018. Accessed: Jun. 20, 2021. [Online]. Available: https://eta-publications.lbl.gov/sites/default/files/feur_10_transportation_electrification_final_20180813.pdf

[12] C. Boll, "As Consumers Focus On Electric Vehicles, Municipalities Look to Electrify Their Bus Systems," *Natl. Law Rev.*, vol. XI, no. 168, Jun. 2021, Accessed: Jun. 20, 2021. [Online]. Available: https://www.natlawreview.com/article/consumers-focus-electric-vehicles-municipalities-look-to-electrify-their-bus-systems

[13] C. Spencer, "Finally, here's the exact cost of owning an electric car vs gas car," *The Hill*, Jun. 23, 2021. Accessed: Jul. 18, 2021. [Online]. Available: https://thehill.com/changing-america/sustainability/energy/559971-finally-heres-the-exact-cost-of-owning-an-electric-car
[14] University of Oxford, "Share of Electricity Production by Source, United States," *Our World in Data*. https://ourworldindata.org/grapher/share-elec-by-source?time=2010..2020&country=~USA (accessed Jun. 21, 2021).
[15] L. Aramayo, "More than 100 coal-fired plants have been replaced or converted to natural gas since 2011," *Today in Energy*, Aug. 05, 2020. Accessed: Jun. 21, 2021. [Online]. Available: https://www.eia.gov/todayinenergy/detail.php?id=44636
[16] A. Lawson, "Variable Renewable Energy: An Introduction," Congressional Research Service, R42455, Jun. 2019. Accessed: Jun. 21, 2021. [Online]. Available: https://crsreports.congress.gov/product/pdf/IF/IF11257
[17] K. Lydersen, "Grid congestion a growing barrier for wind, solar developers in MISO territory," *Energy News Network*, Sep. 29, 2020. https://energynews.us/2020/09/29/grid-congestion-a-growing-barrier-for-wind-solar-developers-in-miso-territory/ (accessed Jul. 11, 2021).
[18] A. MacDonald, C. Clack, A. Alexander, A. Dunbar, J. Wilczak, and Y. Xie, "Future cost-competitive electricity systems and their impact on US CO2 emissions," *Nat. Clim. Change*, vol. 6, pp. 526–531, Jan. 2016.
[19] "How Much Power is 1 Gigawatt?," *Office of Energy Efficiency & Renewable Energy*, Aug. 12, 2019. https://www.energy.gov/eere/articles/how-much-power-1-gigawatt
[20] D. Roberts, "We've been talking about a national grid for years. It might be time to do it.," *Vox*, Aug. 03, 2018. https://www.vox.com/energy-and-environment/2018/8/3/17638246/national-energy-grid-renewables-transmission (accessed Jun. 23, 2021).
[21] S. Borenstein and J. Bushnell, "The U.S. Electricity Industry After 20 Years of Restructuring," National Bureau of Economic Research, NBER Working Paper No. 21113, 2015. [Online]. Available: https://www.nber.org/papers/w21113
[22] J. Hein, D. Hulburt, M. Milligan, L. Coles, and B. Green, "Transmission Planning Process and Opportunities for Utility-Scale Solar Engagement within the Western Electricity Coordinating Council (WECC)," National Renewable Energy Lab (NREL), Technical Report, 2011. Accessed: Jul. 01, 2021. [Online]. Available: https://www.nrel.gov/docs/fy12osti/51279.pdf
[23] J. Stanek, A. Rickert, W. Ellertson, M. Martin, and R. Kessler, *Examining the State of Electric Transmission Infrastructure: Investment, Planning, Construction, and Alternatives.* 2018. Accessed: Jul. 10, 2021. [Online]. Available: https://docs.house.gov/meetings/IF/IF03/20180510/108283/HHRG-115-IF03-20180510-SD002.pdf
[24] "Order No. 888," *Federal Energy Regulatory Commission*, Aug. 05, 2020. https://www.ferc.gov/industries-data/electric/industry-activities/open-access-transmission-tariff-oatt-reform/history-oatt-reform/order-no-888
[25] "About FERC," *Federal Energy Regulatory Commission*, Jul. 06, 2021. https://www.ferc.gov/what-ferc (accessed Jul. 07, 2021).
[26] "Order No. 1000 - Transmission Planning and Cost Allocation," *Federal Energy Regulatory Commission*, Jun. 17, 2020.

[27] M. Brown and R. Sedano, "Electricity Transmission: A Primer," National Council on Electric Policy, Jun. 2004. [Online]. Available: https://www.energy.gov/sites/prod/files/oeprod/DocumentsandMedia/primer.pdf
[28] B. Plumer, "Pipeline Hack Points to Growing Cybersecurity Risk for Energy System," *The New York Times*, Jun. 08, 2021. Accessed: Aug. 05, 201AD. [Online]. Available: https://www.nytimes.com/2021/05/13/climate/pipeline-ransomware-hack-energy-grid.html
[29] IEA, "Power Systems in Transition," IEA, Paris. [Online]. Available: https://www.iea.org/reports/power-systems-in-transition
[30] Mission Support Center, "Cyber Threat and Vulnerability Analysis of the U.S. Electric Sector," Idaho National Laboratory, Aug. 2016. Accessed: Aug. 06, 2021. [Online]. Available: https://www.energy.gov/sites/prod/files/2017/01/f34/Cyber%20Threat%20and%20Vulnerability%20Analysis%20of%20the%20U.S.%20Electric%20Sector.pdf
[31] "Biden Administration Takes Bold Action to Protect Electricity Operations from Increasing Cyber Threats," *Department of Energy*, Apr. 20, 2021. https://www.energy.gov/articles/biden-administration-takes-bold-action-protect-electricity-operations-increasing-cyber-0 (accessed Jun. 23, 2021).
[32] D. Glickson, "Interconnections Seam Study," *National Renewable Energy Laboratory*. https://www.nrel.gov/analysis/seams.html (accessed Jun. 30, 2021).
[33] A. Bloom *et al.*, "The Value of Increased HVDC Capacity Between Eastern and Western U.S. Grids: The Interconnections Seam Study," National Renewable Energy Lab (NREL), Oct. 2020. [Online]. Available: https://www.nrel.gov/docs/fy21osti/76850.pdf
[34] D. Roberts, "Transmission week: how to start building more big power lines," *Volts*, Jan. 27, 2021. https://www.volts.wtf/p/transmission-week-how-to-start-building (accessed Jun. 25, 2021).
[35] B. Alexander, "The Lesson From DOE's Latest Loan Default? Invest More.," *Forbes*, Aug. 05, 2020. [Online]. Available: https://www.forbes.com/sites/brentanalexander/2020/08/05/the-lesson-from-does-latest-loan-default-invest-more/
[36] Loan Programs Office, "Transmission Loan Guarantees," *Energy.gov*, Apr. 30, 2021. https://www.energy.gov/sites/default/files/2021-04/DOE-LPO_Program_Handout_Transmission-April2021.pdf (accessed Jul. 27, 2021).
[37] C. Courchesne, "What Would Burial of High-Voltage Transmission Lines Cost?," *The Northern Pass Alternatives (Part I): Going Underground*, Nov. 17, 2013. https://www.clf.org/blog/going-underground-northern-pass-alternatives-part/
[38] Department of Energy, "DOE Announces Up to $8.25 Billion in Loans to Enhance Electrical Transmission Nationwide," *Energy.gov*, Apr. 27, 2021. https://www.energy.gov/articles/doe-announces-825-billion-loans-enhance-electrical-transmission-nationwide (accessed Jul. 27, 2021).
[39] P. DeFazio, *Infrastructure Investment and Jobs Act*. 2021. Accessed: Oct. 03, 2021. [Online]. Available: https://www.congress.gov/bill/117th-congress/house-bill/3684/text
[40] United States Department of Transportation, "TIFIA Program Overview," *Trasportation.gov*, Jul. 14, 2021. https://www.transportation.gov/buildamerica/financing/tifia (accessed Jul. 27, 2021).

MITIGATION OF PFAS IN PUBLIC WATER SYSTEMS

Future Steps for Ensuring Safer Drinking Water

Alexis Voulgaropoulos
2021 AIChE WISE Intern

AIChE
The Global Home of Chemical Engineers

WISE
WASHINGTON INTERNSHIPS for STUDENTS of ENGINEERING

Preface

About the Author

Alexis Voulgaropoulos is a senior in Chemical and Biomolecular Engineering at North Carolina State University (NCSU) in Raleigh, NC. Throughout college, Alexis has completed four internship experiences within the pharmaceutical industry. After exploring the development of drugs, she decided to investigate the correlation between health, environmental factors, and policy through the WISE program. At her university, Alexis serves as the Professional Development Chair for NCSU's American Institute of Chemical Engineers (AIChE) chapter. She has done academic research for three years under Dr. Albert Keung in creating *in vitro* models of the brain. Additionally, she has served for four years as a College of Engineering Ambassador. After obtaining her Bachelor's, Alexis hopes to pursue a Ph.D. in Chemical Engineering. Long-term, she hopes to work with policymakers to create protective environmental and health regulations.

About WISE

The Washington Internships for Students of Engineering (WISE) is an internship program that was founded in 1980. WISE is an interdisciplinary program that enables engineering students from multiple professional engineering societies, including AIChE, to focus on a public policy issue for 9 weeks over the summer. Interns are tasked with independently writing, researching, and presenting a paper on a public policy issue that is relevant to engineering. Additionally, interns learn about the legislative and decision-making process for complex issues.

Acknowledgements

This paper and internship experience would not have been possible without the help of significant people in my life. First, I would like to thank my mentor Dr. Dale Keairns, for helping me in deciding my topic and guiding my thinking process. I would also like to thank the Faculty Member-in-Residence, Mark Ames, for providing guidance to the other interns and I throughout the course of the summer. Thank you to Heather Yuengling for the AIChE correspondence and connections. I would also like to thank the following:

- My parents and sister, for constantly supporting me no matter what.
- Chris Moody, *AWWA,* Laura Berkey-Ames, *BASF,* Dr. George Gardnier, *EPA,* Sharon Cooperstein, *EPA,* Katherine Hinnant, *NRL*, and Dr. Thomas Speth, *EPA*, for providing me with insights about PFAS, environmental regulations, and public water systems.
- My roommates in Boston, Nikki, Brooke, and Maddie, for encouraging me to pursue this program and helping me talk through my ideas.
- Sarah, Jeremy, and others for giving feedback on my paper.
- Aitana, for supporting me and talking me through my various thought processes.
- Dr. Kim Roberts, for answering all of my questions and believing in me.
- Dr. Lisa Bullard, for introducing me to the WISE program three years ago.
- Dr. Albert Keung, for giving me the confidence to work on independent projects like this.
- The fellow WISE interns, for being great friends and support throughout the summer.

Table of Contents

List of Abbreviations

AC	Advisory Circular
AFFF	aqueous film forming foam
AIChE	American Institute of Chemical Engineers
DaRLP	PFAS Detection and Restriction in Landfills Program
DOD	Department of Defense
DWSRF	Drinking Water State Revolving Fund
EHS	Environmental, Health and Safety
EPA	Environmental Protection Agency
FAA	Federal Aviation Administration
GAC	granular activated carbon
HAZMAT	hazardous materials
MCL	maximum contaminant level
MIL-SPEC	United States Military Specification
MSWLFs	municipal solid waste landfills
NCSU	North Carolina State University
NDAA	National Defense Authorization Act
NIEHS	National Institute of Environmental Health Sciences
NPDWR	National Primary Drinking Water Regulations
NRL	Naval Research Laboratory
PFAS	per- and polyfluoroalkyl substances
PFBS	perfluorobutanesulfonic acid
PFCs	perfluorinated compounds
PFHpA	perfluoroheptanoic acid
PFHxS	perfluorohexanesulfonic acid
PFNA	perfluorononanoic acid
PFOA	perfluorooctanoic acid
PFOS	perfluorooctanesulfonic acid
PFOSA	perfluorooctanesulfonamide
ppt	parts per trillion
PTFE	polytetrafluoroethylene
SDWA	Safe Drinking Water Act
U.S.	United States
UCMR	Unregulated Contaminant Monitoring Rule
USS	United States Ship
WISE	Washington Internships for Students of Engineering
WWTPs	wastewater treatment plants

1 Executive Summary

The paper focuses on two different areas: public water systems and per- and polyfluoroalkyl substances (PFAS). PFAS are emerging chemical contaminants that are persistent, bioaccumulative, toxic, mobile, and thus take a long time to break down in the environment. PFAS were initially created in 1938, have been widely produced by the chemical manufacturing industry since the 1950s. The circulation and usage of PFAS in the consumer market and for essential use purposes have led to PFAS being found in organisms worldwide. With respect to the United States, literature has shown a widespread presence of PFAS in public water systems across the country. The main pollution sources that contribute to PFAS in water are the chemical manufacturing industry, aqueous film-forming foam runoff, and landfills. In the past 20 years, there has been increased concern about PFAS due to recent studies on PFAS toxicity and health effects. Multiple toxicology and epidemiology studies have shown that PFAS can cause adverse health effects, including kidney cancer, testicular cancer, and ulcerative colitis.

Due to the occurrence of PFAS in United States public water systems and the health effects associated with PFAS exposure, action is needed. Various measures have already been enacted. In the 2000s, both PFOA and PFOS were voluntarily phased out by United States manufacturers. In 2016, the Environmental Protection Agency (EPA) set a Lifetime Drinking Water Health for PFOA and PFOS to serve as a non-enforceable guidance to drinking water stakeholders. Currently, bills have been introduced and are being reviewed in the House of Representatives and the Senate. However, further action must be taken to tackle the presence of PFAS in drinking water. The actions should tackle two areas: 1) reducing the contamination at the pollution source and 2) improving quality of PFAS contaminated drinking water. Five recommendations are provided to be reviewed by policymakers and environmental agencies, state-wide and nationally.

1. Individual states should enforce maximum contaminant levels for measurable PFAS in their drinking water on both an immediate and long-term basis.
2. The waste streams of chemical plants involved with PFAS must be regulated and heavily monitored. This involves regulation of PFAS in stack emissions, management of existent PFAS in landfills, prevention of future PFAS landfill leachate contamination, and initiation of PFAS task forces for all chemical plants that use PFAS in their processes.
3. There should be more regulation focused on reducing aqueous film-forming foam in drinking water. An extensive portfolio of effective fluorine-free foams should be developed, and PFAS-containing foam should be phased out of use.
4. Although stopping contamination at its source can reduce PFAS in drinking water, PFAS is still present in drinking water. To ensure safe drinking water, water treatment plants should be equipped with adequate PFAS treatment technologies. Three technologies to consider are granular activated carbon, reverse osmosis filters, and ion-exchange resins. For areas with high concentrations of PFAS in water and inadequate water infrastructure, bottled water and household filters should be provided.
5. To significantly reduce the future presence of PFAS in drinking water, a large-scale "Nothing but the Essentials" ban on PFAS should be introduced. PFAS-free substitutes without the undesirable characteristics of PFAS – persistent, mobile, bioaccumulative, and toxic – should be developed and introduced into the market.

By reviewing and applying these recommendations, there could be a reduction in the release of PFAS into drinking water sources. This would reduce environmental exposure to PFAS, and better ensure safer drinking water for communities across the United States.

2 Introduction

2.1 Defining the Issue: An Overview of PFAS

From their origin in 1938, per- and polyfluoroalkyl substances (PFAS), mainly perfluorooctanoic acid (PFOA) and perfluorooctanesulfonic acid (PFOS), have been used in many products for their surfactant properties. From the 1950s, companies like DuPont and 3M have produced various PFAS products. Over the years, products with these compounds circulated into the commerce of the United States. These products included Teflon non-stick pans, non-stick parchment paper, waterproof materials, floss, and many others [1]. In addition to consumers, PFAS were also used in a fire fighting agent called aqueous film forming foam (AFFF), a chemical mixture used to extinguish oil and chemical-based fires. For more than half a century, consumers across the United States utilized these products without concern. AFFF helped extinguish chemical fires in various military bases, airports, petrochemical plants, and firefighting incidents. Chemical manufacturers used the two main PFAS, PFOA and PFOS, in their production process and disposed of them without regulatory oversight.

However, in 1999, a landmark lawsuit from a community in Parkersburg, West Virginia, began to cast doubt on the safety of these perfluorinated compounds. A cattle farmer from West Virginia had noticed that many of his livestock, roughly 153 cows, were dying and getting sick after drinking the creek water near his farm [2]. The creek was downstream of a landfill owned by DuPont's Washington Works facility, where the chemical company disposed of their PFOA, alternatively known as C8, waste [3]. It was suspected that DuPont's PFOA waste from Teflon production was the reason for these anomalies among the cattle population. The farmer, William Tennant, hired an environmental lawyer to investigate the issue and filed a lawsuit against DuPont in summer 1999. The investigation revealed many issues with the PFOA compound. Past corporate documents showed that the two producers of PFOA, 3M and DuPont, had conducted hidden animal studies using PFOA and observed that the chemical had caused negative health effects, such as birth defects in rats and enlarged livers in rats, rabbits, and dogs [2]. The lawyer, Robert Bilott, published a public brief with his findings against DuPont in March 2001 [2]. This brief led to the EPA giving DuPont a civil penalty of $16.5 million [2].

Although this initial incident affected cattle, it shed light on potential issues that may arise from PFAS. There are multiple entryways for PFAS to enter drinking water sources and, subsequently, public water systems. The lawsuit also resulted in a drastic increase in EPA regulation of the PFOA and PFOS. To accommodate these issues, the primary chemical manufacturers that used PFOA and PFOS voluntarily phased out PFOA and PFOS from their products and manufacturing process in the 2000s. Although these companies took voluntary action to remove PFOA and PFOS from their product line, there remained a larger issue: 50 years of PFAS had accumulated in the environment and public water systems of the United States.

Further, the Parkersburg lawsuit and the voluntary actions brought greater attention to the class of perfluorinated compounds. As more studies and investigations were conducted on these man-made chemicals, more revelations about these surfactants came to light. The following section will introduce the basics of PFAS and cover a comprehensive review of the science and research on PFAS from the past two decades. Additionally, it will review the concerns associated with PFAS, and why it poses an issue for drinking water and human health.

3 Background

3.1 The Chemistry Behind PFAS

PFAS is an acronym for per- and polyfluoroalkyl substances. PFAS are also referred to as perfluorinated compounds (PFCs) [4]. PFAS are a group of man-made chemicals created in 1938. Generally, PFAS are made up of a carbon chain with fluorine atoms attached [5]. Perfluoroalkyl substances, such as perfluorooctanoic acid (PFOA), consist of a fluorinated carbon chain with a functional group attached to the last carbon [6]. Polyfluoroalkyl substances, such as polytetrafluoroethylene (PTFE), have at least one carbon attached to a fluorine [6].

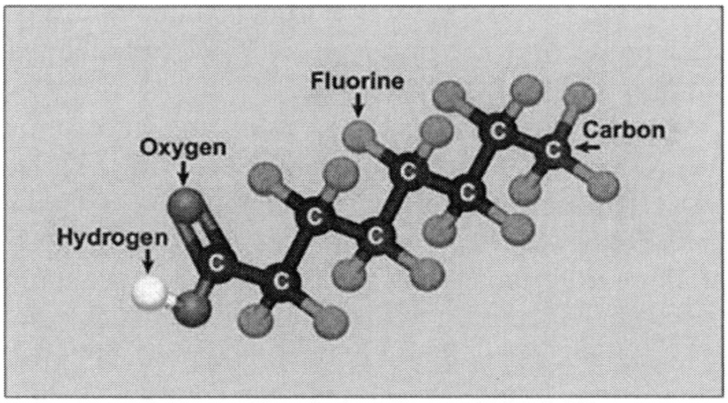

Figure 1. Chemical makeup of PFAS [5]

PFAS have various properties due to their chemical makeup. Primarily, PFAS can act as surfactants, making them water-repellent and oil-repellent [7]. Additionally, due to the stable carbon-fluorine chemical bond, PFAS are chemically nonreactive and heat resistant [7]. For consumer products, these properties aid in water resistance, grease resistance, soil resistance, and making surfaces nonstick. [7]. From their discovery in 1938 to present day, PFAS have been widely integrated into consumer markets. Several products that contain PFAS with these properties are the following [8]:

- Oil-resistance: grease-resistant paper, fast food wrappers/containers, microwave popcorn bags, pizza boxes, candy wrappers
- Stain resistant coating (for several fabrics, carpets, and upholstery)
- Water-resistant coatings
- Cleaning products
- Personal care products and cosmetics
- Paints, varnishes, and sealants
- Aqueous film-forming foams

In addition to these desirable properties, PFAS have characteristics that give reason for concern. The carbon-fluorine bond in PFAS is among the shortest and strongest of known chemical bonds. Due to the strong bond, PFAS are persistent, as they are not easily broken down and remain in the environment for a long time [9][10]. Smaller, short-chain PFAS have increased mobility, allowing them to travel long distances through the environment [11]. The mobility and persistence of the chemicals cause several PFAS to be bioaccumulative, meaning they accumulate in organisms over time [11][12]. PFAS have also recently been found to be toxic and cause adverse health effects, which is further discussed in Section 3.5.

3.2 PFAS Historical Perspective

To adequately understand the issue of PFAS in drinking water, a historical perspective of the origin and use of PFAS must be reviewed. In 1938, a fluoropolymer chemical called polytetrafluoroethylene (PTFE) was discovered by Roy J. Plunkett, a chemist who worked at DuPont. The chemical had various useful properties, as it was chemically stable, non-corrosive, and had a very high melting point [13]. Most prominently, PTFE was low friction, hence its "slippery" and non-stick nature. The fluoropolymer was patented and registered as DuPont's Teflon product in 1945 [14]. In 1946, Teflon was introduced into the consumer market and advertised for its non-stick properties. The discovery of PTFE initiated the creation, production, and use of other per- and polyfluoroalkyl substances.

In the late 1940s, both DuPont and 3M started producing large amounts of PFAS materials. 3M began its large-scale production of perfluorooctanoic acid (PFOA). DuPont produced over 900 tons of PTFE-based Teflon per year at their Washington Works facility in Parkersburg, West Virginia. In the 1950s and 1960s, two PFAS compounds, PFOA and PFOS, started to be implemented into the production of protective coatings, water-resistant products, and stain-resistant products. These products included Scotchgard and non-stick commercial cookware made from Teflon [15]. Outside of the realm of commercial products, PFAS was also used to produce firefighting technologies. In the 1960s, the Naval Research Laboratory (NRL) and 3M developed an aqueous film-forming foam to extinguish large petroleum fires. Following the 1967 USS Forrestal fire in North Vietnam which caused 134 casualties, the Navy mandated that AFFF was carried on all vessels to ensure fire safety [16].

The continuous use and production of PFAS led to the eventual contamination of water sources across the United States [17]. A past example is 3M's Cottage Grove plant in Minnesota, which produced PFAS [18]. From its production start in the 1950s to the 1970s, 3M's Cottage Grove plant legally dumped perfluorochemical waste in landfills near Lake Elmo of Minnesota [19][18]. This resulted in the PFAS leaching into the groundwater and contaminating the community water supply [19]. In 1998, a farmer in Parkersburg, West Virginia, reported the death of multiple cattle and suspected DuPont's nearby Washington Works plant, which produced PFOA, as the source [2]. The lawsuit found that 7,100 tons of highly-concentrated PFOA waste had been dumped by DuPont into a nearby landfill [2]. It was confirmed that the Washington Works plant released contaminants into nearby water sources [2].

Moving into the 21st century, greater attention was brought to the potential environmental and health effects of the two primary perfluorinated compounds used in the supply chain: PFOA and PFOS. In 2001, an article showed the widespread presence of PFOS in wildlife worldwide, illustrating the persistent and bioaccumulative nature of PFOS [20]. This article raised international concern about PFAS and its effects [1]. As a result, 3M vouched to voluntarily phase out PFOA from production by 2002 [32]. By 2006, the EPA's 2010/2015 PFOA Stewardship Program was initiated, setting a goal to eliminate PFOA emissions from 8 companies in the United States [21]. A 2007 study found that four PFAS were detected in the blood of 98% of the U.S. population [22].

Despite the phaseouts of two legacy, long-chain PFAS, PFOA and PFOS, in the United States, these two chemicals can still enter the consumer economy through internationally imported products [23]. Additionally, due to the phaseouts of the legacy PFAS, short-chain

PFAS alternatives have been implemented into products and manufacturing processes. Although these shorter chain "replacement" PFAS serve as an alternative, recent studies have shown that these non-legacy chemicals have similar properties to their longer chain constituents and may pose an even greater threat to human health [24].

The history of PFAS thus far shows the emergence of the chemicals in modern consumer markets and lifestyles. Additionally, it demonstrates the rising concern associated with both legacy and replacement PFAS, giving a glimpse at the need for future policy action.

3.3 Public Water Systems

The scope of the paper focuses on PFAS presence in public water systems, as these systems provide drinking water to a large number of Americans. A public water system is a pipe-based water system that provides drinking water to at least 25 people for at least 60 days per year or has "at least 15 service connections" [25]. Over 148,000 public water systems in the United States serve approximately 317 million consumers [25][26]. There are multiple types of water systems where people access drinking water. Community water systems, the most common type, provide tap water for drinking to over 286 million people [27]. Both surface water and groundwater are important sources for providing drinking water to people in the United States, as community water systems get their water from these sources [27].

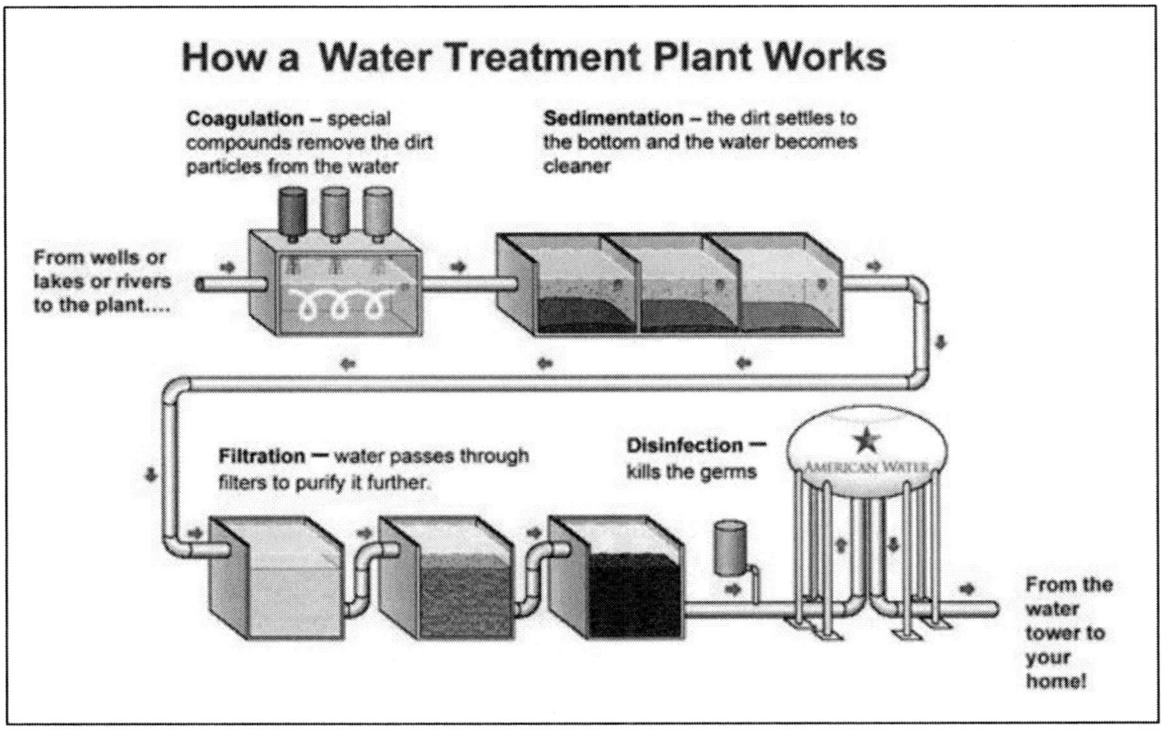

Figure 2. Illustration of the Water Treatment Process [28]

To ensure drinking water is potable, the water collected from these sources must be treated at water treatment plants. As the water is pumped into the plant, various floating solids, such as wood, leaves, and other large debris, are filtered out using screening [29]. Then, sediment and organic material are removed through coagulation and flocculation. In coagulation, chemicals called coagulants are added to the water, neutralizing the charges in the particles and allowing them to stick together. Once the particles are aggregated, the water is transferred to a flocculation basin. In the basin, slow mixing causes the formation of floc particles, which are clumps of the particles. The water is then transferred to sedimentation basins, where the floc particles accumulate at the bottom of the basin due to gravity. These gathered floc particles become sludge. The sludge is normally taken to a landfill or holding facility [30].

After larger particles are removed, the water is moved to filtration for the removal of remaining particles and microbes. Filtration occurs by running the water through layers of fine, granulated particles, using either sand, gravel, or carbon beds as filters [29][31]. Finally, disinfection occurs, where chlorine compounds are used to eliminate microorganisms from the water. Additionally, some plants may add alkaline substances to the water to maintain its pH and prevent corrosion of pipes [31]. After all of these steps, water is either pumped and distributed through the city's pipe distribution system, or stored for future use [30].

Water treatment plants in public water systems must comply with standards set by the Safe Drinking Water Act (SDWA) [32]. The current drinking water regulations are listed under the National Primary Drinking Water Regulations (NPDWR), which is governed by the Safe Drinking Water Act [33]. These regulations have enforceable limits, called maximum contaminant levels (MCLs), for over 90 contaminants that can be found in drinking water [33].

3.4 The Safe Drinking Water Act

In 1974, the EPA enacted the Safe Drinking Water Act [34]. The purpose of the law was to "protect public health regulating the nation's public drinking water supply" [35]. Over time, the law has been amended to have greater coverage of public drinking water systems. More prominently, the SDWA enables the EPA to make national drinking water standards to mitigate health risks associated with contaminants [36]. These drinking water standards require water systems to test for the contaminant of interest and ensure that it is at or below the set limit [35]. The SDWA targets four types of potential contaminants to drinking water: physical contaminants, chemical contaminants, biological contaminants, and radiological contaminants [37]. If these contaminants are not adequately managed and treated in a public water system, the contaminants can pose a health risk to humans.

Through evaluation of potential contaminants of concern, their associated health risks, and their frequency of occurrence, the EPA can decide whether to instill a national drinking water regulation for a contaminant [38]. For this to occur, the contaminant must have potential for causing a health issue, have a frequent occurrence in multiple public water systems, and regulatory action must have potential for reducing health risks associated with the contaminant [38]. Currently, PFAS are not regulated by the National Primary Drinking Water Regulations.

3.5 Health Effects of PFAS

Multiple epidemiology and toxicology studies have been conducted on the health effects associated with PFAS. These studies have been primarily funded by the National Institute of Environmental Health Sciences (NIEHS).

The National Toxicology Program conducted multiple rat toxicology studies on the potential health effects of PFAS. A 28-day study found that both long- and short-chain PFAS affect liver and thyroid hormones [39]. Another study illustrated increased numbers of pancreatic tumors in male and female rats and increased number of liver tumors in male rats for rats exposed to PFAS during gestation or post-weaning [39]. PFAS impacted and changed mitochondrial function, with long-chain PFAS having more effect on mitochondrial function than short-chain PFAS. In a neurodevelopment study, it was observed that certain PFAS (PFOS, PFBS, and PFOSA) had potential to impact neurodevelopment due to modification of neural cell differentiation [40]. These studies suggest that high doses of exposure to certain PFAS can cause health issues, however, due to differences between mouse models and humans, these health risks from the rat studies are not conclusive for human health risk.

Due to the large amount of attention given to PFAS after the voluntarily PFOA and PFOS phase-out, many academic health effect studies were initiated in the 2000s. A large amount of current human health data for PFAS, specifically PFOA, has been obtained from the C8 Health Project, which studied the affected population that lived in vicinity of Dupont's Washington Works facility in West Virginia [1]. This population's drinking water was contaminated with PFOA from 1984-2004 [41]. **The C8 Health Projects found a positive association between PFOA exposure levels and testicular and kidney cancer, suggesting possible carcinogenicity associated with PFOA [1]. Additionally, the study found other potential health effects associated with PFOA exposure: pregnancy-induced hypertension, high cholesterol, ulcerative colitis, and thyroid disease [1]. Various other studies have shown an increased occurrence of bladder cancer, kidney cancer, prostate cancer, liver cancer, leukemia, and kidney disease [15].** The Agency for Toxic Substances and Disease Registry also lists two other potential health effects related to birth and pregnancy: small decreases in birth weights and pre-eclampsia in pregnant women [42].

Although research is still being conducted on health effects caused by PFAS exposure, the current studies show a reason for concern, as there are strong correlations between various PFAS and adverse health effects. One of the main pathways of human exposure to PFAS is drinking water [1].

4 Key Concerns – Drinking Water Contamination

The occurrence and persistence of PFAS in public water systems in the United States give rise to three main areas of concern. First, there are many existing and accumulated PFAS in public water systems, affecting drinking water. Second, PFAS-containing compounds are still produced and existent in United States commerce and being emitted into the environment. Third, emerging studies point to many potential adverse health effects of the PFAS in these existent contaminated sites and incoming sources of PFAS. This combination of factors signifies a public health urgency to develop policy strategies that will remove existing PFAS from public water systems and prevent future PFAS contamination of public water systems.

4.1 Contamination of PFAS in the Public Water System

Contamination has been occurring in the public water systems since the 1950s. PFAS are soluble in water, making it easy for them to dissolve in drinking water sources [43]. **The main PFAS contamination sources are industry, run-off from air force bases and airports, fire training, and landfills** [1].

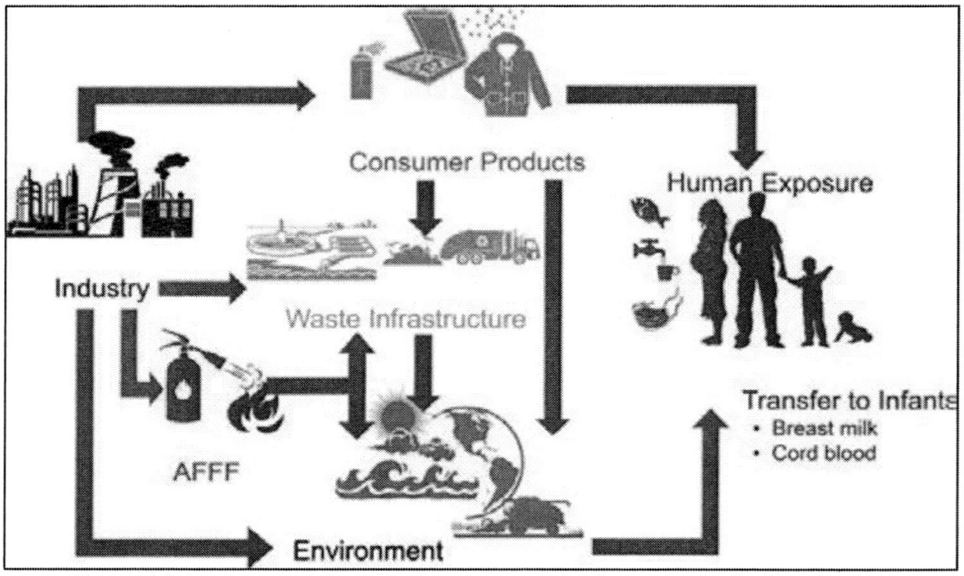

Figure 3. The multiple pathways that PFAS can enter drinking water and affect human health [1]

As shown in Figure 3, PFAS-containing consumer products are an additional source of PFAS contamination in humans [1]. Due to the scope of this paper, only the impact of PFAS-containing products on drinking water systems will be discussed.

Figure 4 showcases the contamination occurrence of PFAS in different locations across the United States and the associated pollution source.

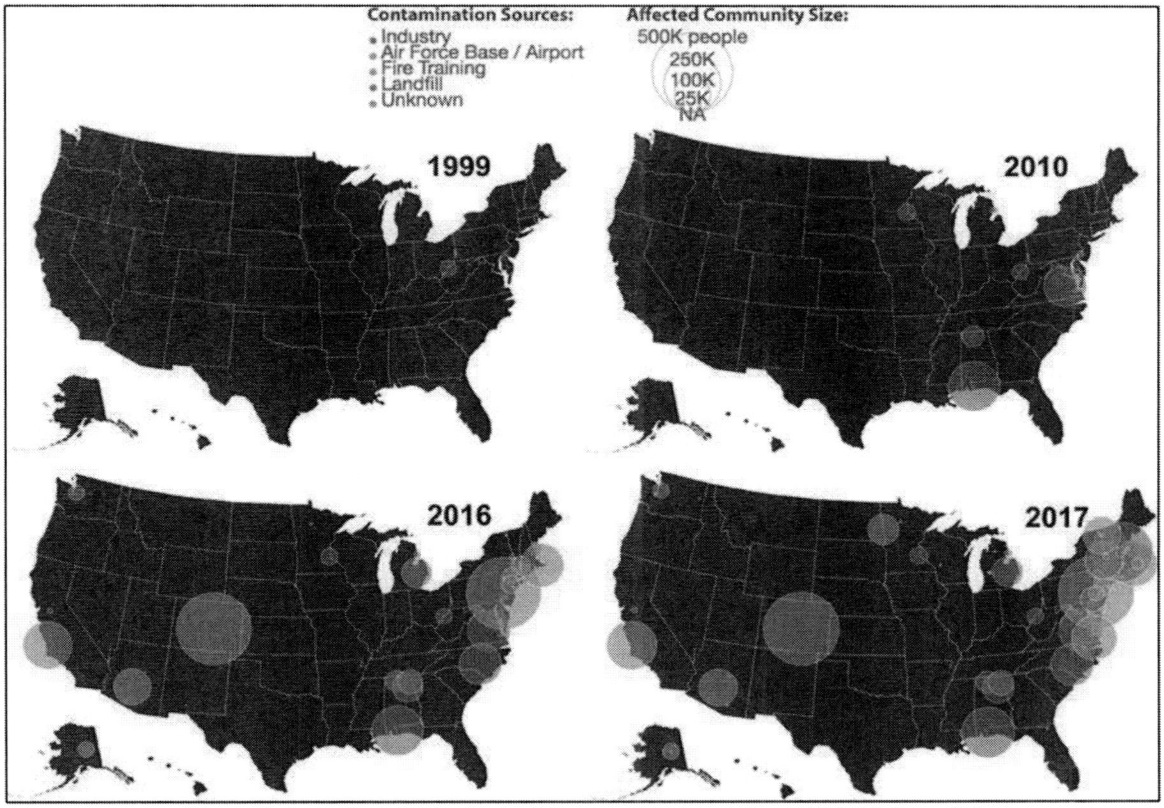

Figure 4. Locations in the United States with contaminated PFAS Sites [1]

From 1999 to 2017, there was an increasing number of PFAS-contaminated sites among all four contamination sources. The two most prominent sources were industry, specifically chemical manufacturing plants, and bi-products of using AFFF at air force bases, airports, and in fire training [1]. These plants and AFFF-applied sites contribute to the presence of PFAS in waste management systems due to lack of containment and proper disposal.

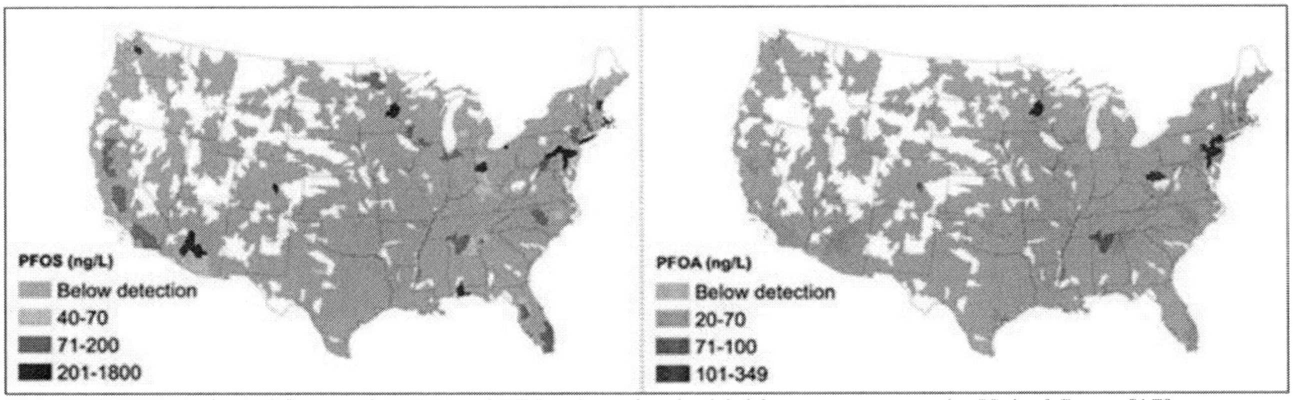

Figure 5. Adapted figure of PFOA and PFOS detection in drinking water across the United States [17]

Additionally, a 2016 EPA document showed that over 75% of detected PFAS were found in 13 different states: California, New Jersey, North Carolina, Alabama, Florida, Pennsylvania, Ohio, New York, Georgia, Minnesota, Arizona, Massachusetts, and Illinois [17]. Figure 5 shows the levels of detection across the United States for PFOA and PFOS, illustrating the widespread occurrence of PFAS in drinking water across the United States. The following section will address how the PFAS has gotten into drinking water, and characterize the pathways.

4.2 Industry Contamination

Industry is a significant source of PFAS in drinking water [17]. The industrial category consists of industrial facilities that use PFAS in their manufacturing, produce PFAS, or process PFAS [44]. Over the course of 70 years, industry has contributed greatly to PFAS pollution into groundwater and drinking water. Studies have pointed to increased concentrations of PFAS in drinking water for regions with fluorochemical-related sites nearby [17]. Specifically, every additional industrial site within an area was associated with an 81% increase of PFOA levels, one of the legacy long-chain PFAS, in drinking water [17]. Figure 6 visualizes the association between specific industrial sites and the detection of six different fluorochemical compounds [17]. The industrial sites are sourced from EPA's 2010/2015 Stewardship program (see Appendix Table 1). The PFAS detection data originates from the EPA's Unregulated Contaminant Monitoring Rule (UCMR), also discussed in Section 4.5.2.

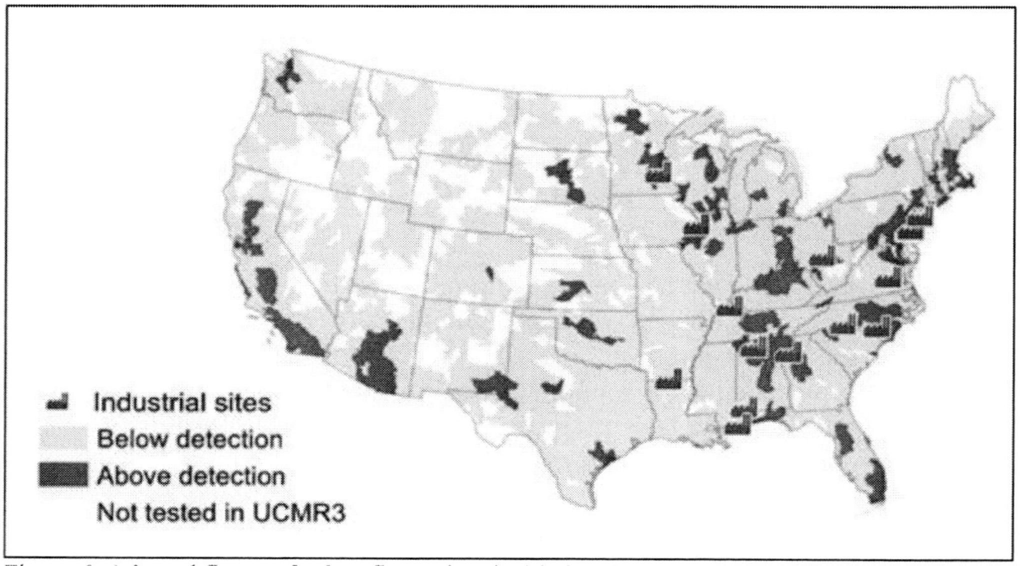

Figure 6. Adapted figure of select fluorochemical industrial sites and PFAS detection data [17]

There are multiple pathways in which PFAS can enter drinking water systems from industrial sources. The primary pathway of PFAS release is through waste disposal. First, the chemicals can enter the water system through **liquid waste disposal in wastewater and stormwater discharge**. This shifts the PFAS contamination to secondary contamination sources: wastewater treatment plants and waste infrastructure. Additionally, PFAS in disposed solid waste, such as **treatment sludge in landfills**, can contribute to contamination of groundwater through soil infiltration and surface water through runoff. The third pathway of PFAS contamination by industry is **accidental leaks and spills** [44]. The final pathway is from **stack emissions**. A stack is an industrial chimney that releases gas emissions and disperse harmful pollutants over a broad area [45]. PFAS gas emissions released from industrial stacks can potentially result in the deposition of PFAS particles on land and surface water [24]. Figure 7 further illustrates the pathways in which PFAS enters water sources from industry.

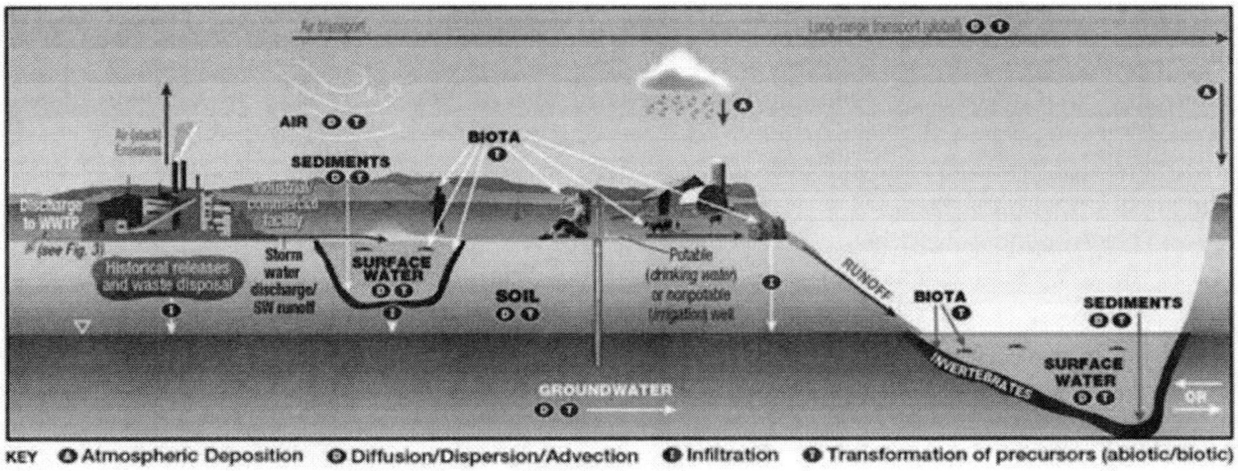

Figure 7. Adapted figure of a Conceptual Site Model for the industrial pathways of PFAS contamination [43]

4.3 PFAS in Waste Management Infrastructure

Due to both industry and consumer actions, PFAS can enter drinking water through various waste management systems. There are multiple pathways in which PFAS can go into waste infrastructure: landfills, land application of biosolids, and wastewater treatment plants.

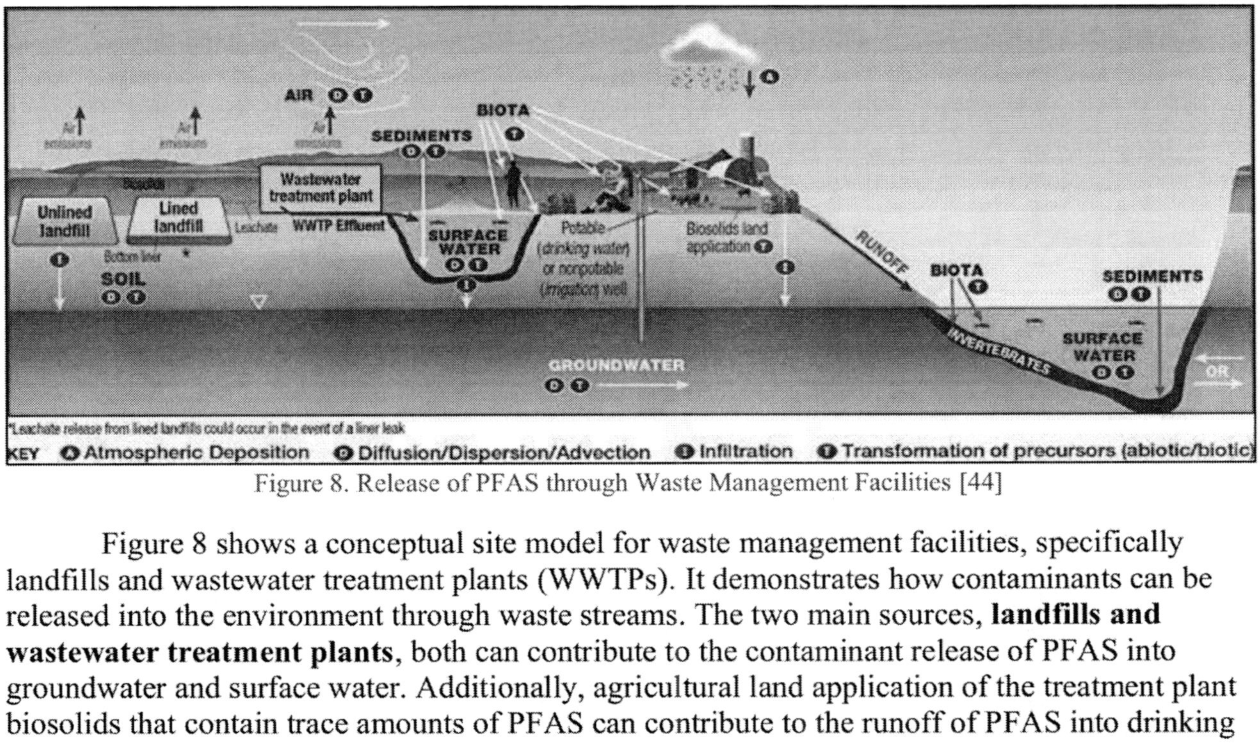

Figure 8. Release of PFAS through Waste Management Facilities [44]

Figure 8 shows a conceptual site model for waste management facilities, specifically landfills and wastewater treatment plants (WWTPs). It demonstrates how contaminants can be released into the environment through waste streams. The two main sources, **landfills and wastewater treatment plants**, both can contribute to the contaminant release of PFAS into groundwater and surface water. Additionally, agricultural land application of the treatment plant biosolids that contain trace amounts of PFAS can contribute to the runoff of PFAS into drinking water.

4.3.1.1 Pathways into Waste Infrastructure: Industry and Consumers

Overall, *industry* is the main contributor to PFAS entering waste infrastructure. Companies that make PFAS-containing products dispose of their manufacturing waste products through various waste disposal systems, mainly **landfills** and **WWTPs**. Past industrial documentation from 3M shows evidence of PFAS waste disposal into landfills and WWTPs [19]. The appendix summarizes 3M's "Fluorochemical Use, Distribution, and Release Overview" report (see Appendix Text 1).

Industry also acts as an indirect contributor, as fluorochemical companies create consumer products with PFAS. These products are eventually used and disposed of by *consumers* and enter landfills. Both PFAS from direct industry disposal and consumer products contribute to the formation of PFAS-containing landfill leachate [46]. Landfill leachate from solid waste landfills is a large contributor of PFAS into water sources [47]. Landfill leachate is a liquid that forms when moisture, such as rainwater, filters through landfill waste and forms highly toxic wastewater [48][49]. Through weather and time, mechanical breakdown of these products occurs, and the PFAS exit landfills as landfill leachate. The liquid PFAS leachate that exits the solid waste landfills enters nearby water sources and becomes contaminated wastewater.

This section conveys that industry is the main source due to its direct contributions through industrial waste and indirect contributions to consumer waste. Action is needed to reduce drinking water contamination from PFAS waste disposal and landfills.

4.4 Aqueous Film-Forming Foam (AFFF) Contamination

In addition to industry directly contributing to PFAS contamination, industry indirectly affects PFAS in drinking water through a key product: aqueous film-forming foam (AFFF). Industry produces AFFF, a life-saving firefighting foam used by fire departments, military sites, and specific airports for training and emergency purposes.

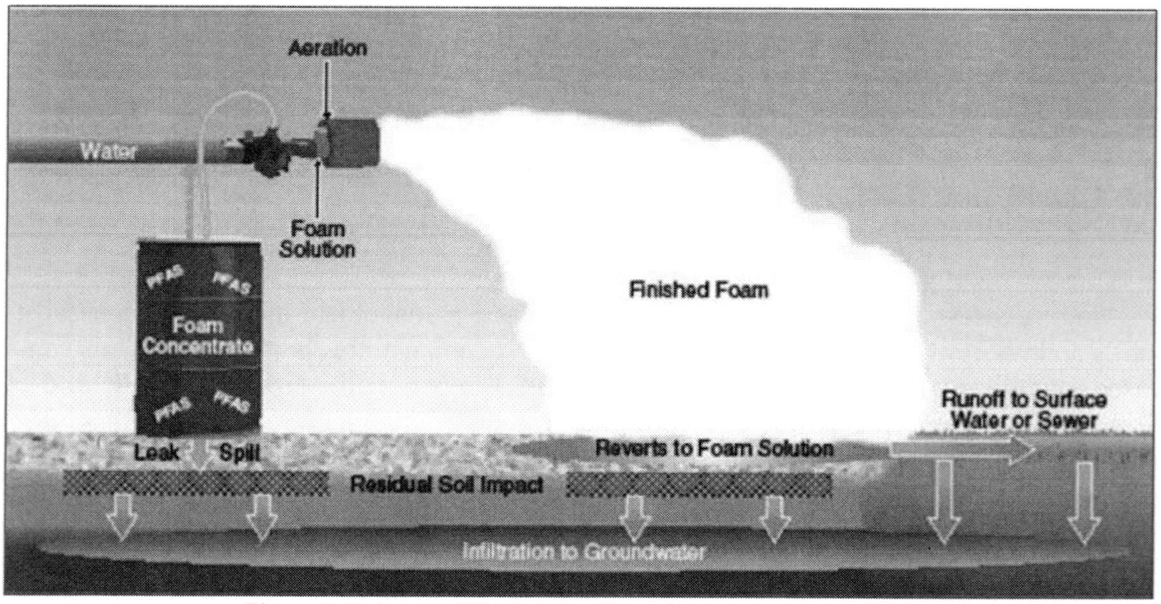

Figure 9. Release of Firefighting Foam into water sources [50]

Class B AFFF is used to suppress hazardous, flammable liquid-based fires, primarily stemming from hydrocarbon fuels [51]. AFFF is used for both firefighting training exercises and for emergency live oil fires [19]. AFFF is used in multiple types of sites: military fire training sites, civilian airports, fire departments, firefighting training facilities, petroleum refineries, fuel storage facilities, and chemical manufacturing plants [51]. In these scenarios, AFFF is sprayed on oil, jet fuel, and petroleum fires to suppress and extinguish the fires in less than 30 seconds [52]. It is used by mixing the foam with water and spraying the foam mixture onto the fire [50]. As shown in Figure 9, PFAS can enter waterways through AFFF runoff [51].

The use of AFFF in fire-training exercises and firefighting activities has resulted in both controlled and uncontrolled releases of the fluorinated foam into groundwater [53]. It was found that the amount of military fire training sites in an area are positively associated with a concentration increase of six different polyfluorinated compounds in drinking water [17]. Markedly, every additional military fire training site in an area accounted for a 35% increase of PFOS levels, a 20% increase of PFHxS levels, and a 10% increase of PFOA levels in drinking water [17]. This suggests that military fire training sites that use PFAS-containing AFFFs are a prominent contributor to the presence and persistence of PFAS in drinking water systems. A future challenge is reducing the release of AFFF from sites that use PFAS-containing AFFF.

AFFF Requirements

Currently, AFFF used in military fire training exercises must adhere to performance requirements of military standard Mil-F-24385F [54]. The Mil-F-24385F performance standards require that AFFF agents extinguish smaller heptane fires ($28ft^2$) within 30 seconds using foam mixed with potable water or seawater. For larger fires ($50ft^2$), the AFFF agent must be able to extinguish the fire within 50 seconds using foam mixed with seawater [52]. Additionally, specifically requires that the firefighting agent is diluted with fresh or seawater at two concentrations: three-part concentrate and 97-part volume water, or six-part concentration and 94-part volume water [52]. The military specification applies to military sites and FAA-regulated airports. Additionally, the Federal Aviation Administration (FAA) has set standards for containing jet fuel-based fires with aircraft fire extinguishing agents. These guidelines are contained within the FAA's Advisory Circular (AC) for Aircraft Fire Extinguishing Agents [54]. The guidelines adhere to the military's Mil-F-24385F.

4.5 Recent PFAS Actions and Plans

Due to the attention on PFAS, various regulations have been implemented in the past decade. Currently, several regulations give guidance on and monitor PFAS in drinking water.

4.5.1 Drinking Water Advisories

Section 3.4 reviewed the Safe Drinking Water Act. The SDWA does not yet enforce the limits of any PFAS in drinking water. However, in 2016, EPA has set Lifetime Drinking Water Health Advisories for both PFOA and PFOS of 70 parts per trillion (ppt) [55]. Currently, there is no maximum contaminant level for PFAS. Federally, PFAS levels in drinking water cannot be enforced and therefore can only be enforced statewide. The advisory is non-enforceable and serves as guidance to state agencies, drinking water system operators, tribal officials, and local officials [55].

4.5.2 Unregulated Contaminant Monitoring Rule

For contaminants that are currently not enforced, such as PFAS, data is collected on the presence of the contaminants in water systems through the Unregulated Contaminant Monitoring Rule (UCMR). The rule is part of the Safe Drinking Water Act (SDWA). UCMR requires water systems operators to test for 30 unregulated contaminants every five years [38]. In 2012, the EPA released UCMR 3, which required approximately 5000 water systems to test for 6 different PFAS (including PFOA and PFOS) between Jan 2013 and Dec 2015 (see Appendix Table 2) [38]. 1.3% of the water systems under this classification found PFOA and/or PFOS content at concentrations above the EPA's health advisory level of 70 ppt [38]. The information gathered in UCMR 3 aims to serve as a basis for "future regulatory actions to protect public health" [56].

In March 2021, the EPA published UCMR 5. UCMR 5 contains 29 perfluorinated compounds [57]. Data will be collected from January 2023 to December 2025 for UCMR 5 [57]. The large presence of PFAS in UCMR 5 signifies the current administration's initiative towards addressing the presence of PFAS in drinking water.

4.5.3 EPA PFAS Action Plan

In 2019, EPA released its PFAS Action Plan. This report was a comprehensive analysis and action plan of future tasks to reduce PFAS in the environment and develop greater analysis and detection tools. One action item of interest is the development PFAS Under the Drinking Water State Revolving Fund (DWSRF) [58]. This fund aims to provide financial aid to drinking water systems, both public and private, that have been affected by PFAS contamination. The fund must be used towards either achieving NPDWR compliance or achieving the health protection objectives of the SDWA [58].

4.6 Problem Statement

As shown by Section 4.1, the detection of PFAS in drinking water has occurred across the United States. Additionally, various adverse health effects are associated with exposure to PFAS. Every day that passes leads to a greater risk of health effects for those exposed to higher concentrations of PFAS. The sources of PFAS in the U.S. drinking water have been largely identified and characterized. Despite occurrence and adverse health effects associated with PFAS, there are no currently enforceable nationwide limits that have been implemented, and PFAS policy is still developing. Therefore, action is needed to remove PFAS from drinking water by controlling pollution sources and developing treatment strategies for existing PFAS.

5 Policy Recommendations

The main issue is that PFAS are currently contaminating public water systems and have contaminated public water systems since the 1950s. Various epidemiology and toxicology studies have shown that this contaminated drinking water consumption can have health risks. There have been many different proposed solutions to this issue. As PFAS in drinking water is a complex issue, it is difficult to spearhead one specific solution. The following section will list policy recommendations for alleviating the issue and investigate the implications of each recommendation. The recommendations fall into two categories: reducing contamination from pollution sources, and improving water quality where PFAS is already present in drinking water.

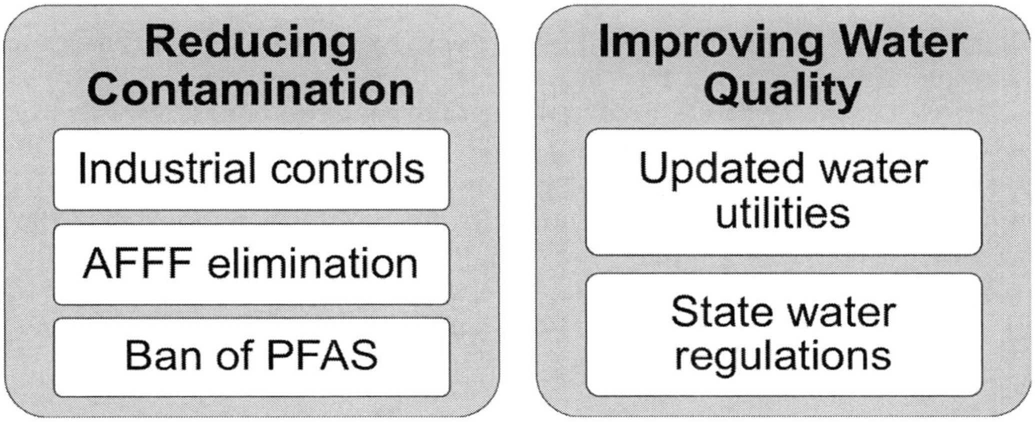

5.1 Ensuring Safer Drinking Water

5.1.1 Drinking Water Regulations

The first action that should be taken is the implementation of enforceable state drinking water regulations for characterized PFAS. Currently, the EPA has a set lifetime drinking water health advisory of 70 ppt for two PFAS: PFOA and PFOS [55]. Although non-enforceable, this technical information from the EPA can serve as a standard for states to make state-enforced drinking water standards for PFAS.

IMMEDIATE ACTIONS:

The Unregulated Contaminant Monitoring Rule (UCMR) acts as an informative document for data of unregulated contaminants potentially present in drinking water [59]. **Although the EPA cannot make limits on contaminants without sufficient data, states and their environmental agencies can make stricter, more protective limits than the EPA.**

From 2013 to 2015, UCMR 3 investigated and measured the occurrence of six PFAS: PFOS, PFOA, PFNA, PFHxS, PFHpA, and PFBS [59]. UCMR 3 showed the detectable amounts of these six PFAS in 4064 public water systems across the United States **[17]. The implementation of MCLs is highly recommended for states with high concentrations of PFAS compounds in their drinking water, based on UCMR 3 [33].** These states are

California, New Jersey, North Carolina, Alabama, Florida, Pennsylvania, Ohio, New York, Georgia, Minnesota, Arizona, Massachusetts, and Illinois [17]. The MCLs should be set and enforced by each state's environmental agency.

Immediate Future Action:

The EPA currently has analytical methods for 29 PFAS: EPA Method 533 and EPA Method 537.1 [57]. UCMR 5 outlines an information summary of the 29 PFAS. Sample collection for UCMR 5 is planned to occur for two years, from January 2023 to December 2025 [60]. After December 2025, once UCMR 5 is published with sample detection data for the 29 PFAS, states that are most affected should develop MCLs for PFAS of highest occurrence. This regulatory action would include the following actions:

1. **Mapping areas with the greatest PFAS release by industry, fire training sites, and corresponding wastewater treatment plants**
    - **Out of the 29 PFAS, occurrence data of each compound will identify the PFAS compounds of concern.**
2. **Setting drinking water standards based on the state's PFAS of concern**
3. **Implementing a state monitoring campaign for quarterly monitoring of the PFAS of concern**
    - **The monitoring would utilize the analytical methods listed in EPA Method 533 and EPA Method 537.1. The monitoring would be conducted by the state's environmental agency.**

Long-term Future actions:

As more analytical methods are developed, this procedure should be repeated for each new PFAS that can be analyzed. New MCLs should be set as more UCMR's are released on a five-year basis.

One obstacle for future long-term actions is that there are still many newer, uncharacterized PFAS with little to no toxicology data being used as substitutes for legacy PFAS. For the EPA and state agencies to act properly, the EPA and NIEHS should establish a basis of research and toxicity classification for PFAS. For this to occur, the House Committee on Energy and Commerce could introduce appropriation bills to increase funding to key agencies, namely the EPA and NIEHS. Increasing funding for PFAS classification purposes allows the EPA and other agencies to better characterize risks associated with a wider range of PFAS used in industry and by consumers.

Setting maximum contaminant levels is only one of the steps that can be taken to alleviate the issue of PFAS in public water systems. Action should also be taken to reduce the source of PFAS entering drinking water. The following sections illustrate how various sources of PFAS can follow steps to reduce the presence of known and measurable PFAS in drinking water.

5.2 Reducing Industry Contamination: Targeted Regulatory Measures to Reduce Industry PFAS Emissions

Section 4.2 introduces one of the main sources of PFAS into drinking water: industry. To reduce the presence of PFAS in drinking water, the release of PFAS by industrial sources should be closely monitored and managed. The three industrial sources of focus are air emissions, landfill release, and leaks and spills.

5.2.1 Regulation of Stack Emissions

Past literature points to stack emissions, which are industrial gas emissions, as a possible contributor to PFAS in water sources [44]. Although more studies need to be conducted on the exact role of industrial gas emissions on water contamination, regulations should be implemented to reduce the release of PFAS from stacks. The following actions should be considered for industrial plants that produce, use, or process PFAS.

For all fluorochemical plants that have high detection of PFAS in nearby drinking water, based on UCMR data, air emissions monitoring should be implemented to measure PFAS in industrial gas emissions. Currently, the EPA has several standard analytical methods for measuring PFAS in stack emissions. The methods listed in the appendix can be used to test for the presence of over 50 PFAS compounds (see Appendix Table 3). The testing is to be managed by the state's environmental agency, the responsible company, and individual contractors hired by the company. If measurable amounts of PFAS at a set level are detected in stack emissions from a facility, regulations in the form of air pollution limits should be enacted and enforced to restrict output of PFAS. The exact air quality regulations are out of the scope of this paper and will not be discussed further. The limits should be applied on a case-by-case basis.

To comply with air emission regulations, it is highly recommended that pollution control technologies are implemented at industrial fluorochemical facilities that use, produce, and process PFAS. For facilities with high amounts of PFAS in emissions, thermal oxidizers could be installed to remove PFAS from the process and effectively dispose of the PFAS in air emissions. Past state legislation has exhibited success in this area. In 2020, after a NC DEQ Consent Order required control of PFAS air emissions, Chemours installed a thermal oxidizer facility. The thermal oxidizer was able to remove over 99.999% of PFAS from air emissions, showing the effectiveness of the technology [61]. Additionally, corresponding air quality permits can be enacted by the state's environmental agency.

Commercial contractors list the cost of thermal oxidizers as ranging from $100,000 to $500,000. Chemours reportedly paid $100 million to implement and operate their thermal oxidizer [62]. Due to the costly nature of large-scale thermal oxidizers, the implementation of thermal oxidizers should only occur if significant pollution has been detected and shown to cause large-scale drinking water contamination. In the interest of contamination prevention, it is suggested that all industrial plants that produce, use, or process PFAS implement wet scrubbers in stacks to reduce emissions of PFAS particles.

5.2.2 Landfill Management

Another source of industry PFAS contamination is PFAS-containing landfill leachate being released and entering the water supply [44]. As of present, PFAS are considered non-hazardous waste. Due to this classification, PFAS from industry and consumers can be disposed of into municipal solid waste landfills (MSWLFs) [63]. For PFAS in landfills, two approaches should be considered.

Managing Existent PFAS In Landfills.

The first approach is managing the already existent PFAS in landfills. As noted, both industry and consumers dispose of PFAS into municipal solid waste landfills. Although many modern landfills have liners and leachate collection systems, PFAS leachate has been found in landfills and the surrounding environment [7]. To prevent further contamination, a PFAS Detection and Restriction in Landfills Program (DaRLP) should be developed for landfills nationwide.

1. As a general precaution, all landfills within a given radius of drinking water sources used by public water systems should be tested for PFAS in soil.
   - The EPA should establish the given radius.
   - The amount and types of PFAS tested will be limited to the approved analytical methods by the EPA. As the EPA develops and approves more analytical methods for PFAS detection, more PFAS can be potentially detected. At present, two emerging analytical methods for soil and other solid matrices are the solid-phase extraction/isotope dilution method and total organic fluoride analyses [64].
2. If PFAS are detected at levels above EPA specification, a remediation and containment plan should be developed.
3. After implementing the plan, the modified landfills should be checked routinely, possibly every 12 months, to ensure minimal leachate leakage.
4. All landfills, affected or not, should be checked at least once within a decade.

These steps are highly recommended to prevent possible future groundwater contamination by leachate. This regulation could be spearheaded by the Senate Environment and Public Works Committee and/or the House Energy and Commerce Committee. If the bill, the responsibilities can be delegated to the EPA, specifically to two offices: the Office of Land and Emergency Management and the Office of Groundwater and Drinking Water. One option for funding of PFAS analytical methods could be a perfluorochemical company tax, where taxes are slightly increased for fluorochemical companies that incorporate PFAS in their processes. If remediation and amendments are recommended for affected PFAS-contaminated landfills, responsible parties could be an option for financing technology implementation.

Prevention of Future Landfill Leachate Contamination

The second approach is the prevention of future drinking water contamination by PFAS landfill leachate. Collected landfill leachate can be treated at wastewater treatment plants. However, there is a lack of effective removal technologies. The most stringent approach to preventing large-scale landfill leachate leakage from municipal solid waste landfills and ineffective removal of PFAS in WWTPs of PFAS is to block disposal of PFAS into MSWLFs in the first place. To safely contain PFAS and avoid any potential leakage, PFAS should be designated as hazardous waste within the next five years. Hazardous waste is waste that can have negative health and environmental effects [65]. Due to the high variety of consumer waste, PFAS embedded in consumer waste products is more difficult to control. Therefore, this hazardous waste designation should only apply to PFAS waste produced by industry. This designation would allow "cradle-to-grave" regulation of the hazardous PFAS waste, from formation to disposal [65]. By disposing of industry PFAS waste streams as hazardous waste, less PFAS will enter MSWLFs, and one pathway of PFAS into drinking water is reduced. Treatment of disposal of the contained hazardous PFAS waste presents a future challenge that should be evaluated by the EPA and key PFAS disposal research organizations.

5.2.3 Leak and Spill Prevention Measures

It is suggested that companies within the PFAS industry have a PFAS task force that implements leak and spill prevention measures. Task forces of this nature could be a part of the facility's Environment, Health and Safety (EHS) department and have the following responsibilities:

- Budget appropriations of PFAS containment technologies
- Routine monitoring of plant waste streams
- Immediate action towards detected PFAS in waste streams
- Collaboration with state and federal environmental agencies to ensure contaminants are being properly managed and surrounding communities are being protected

Industrial facilities that use PFAS in their manufacturing, produce PFAS, or process PFAS are highly recommended to initiate a PFAS task force.

The suggested policy recommendations can reduce the industrial release of PFAS into drinking water sources. Other guidance and recommendations should be explored to get a full view of the issue and potential solutions.

5.3 Reducing Aqueous Film-Forming Foam in Drinking Water

AFFF has been in use since the 1960s and is a critical component of firefighting safety. It was initially produced by 3M and the Naval Research Laboratory [16]. AFFF is used at various sites for extinguishing hydrocarbon fires: petroleum plants, military training facilities, firefighting training, and fire departments [51]. As mentioned in the background, AFFF is one of the main contributors to PFAS in drinking water. The AFFF enters the drinking water through seeping into groundwater and through runoff into surface water [51]. There are multiple avenues for addressing and reducing AFFF PFAS contamination.

5.3.1 "No PFAS in Firefighting Training Foam" Legislation

The National Defense Authorization Act of 2020 (2020 NDAA) prohibits "uncontrolled release" of fluorinated AFFF at military sites starting on October 1, 2024. An uncontrolled release is the "release of a hazardous substance from its container" [66]. The law means that all AFFF used in non-emergency situations must be contained, except for AFFF used in emergency responses [67][51]. For AFFF to be used in training exercises, thorough and proper containment, capture, and disposal mechanisms must be set up to prevent AFFF release into the environment. Additionally, the law allows the use of fluorine-free extinguishing agents in training exercises instead of just AFFF.

2020 NDAA presents a suggestion towards eliminated AFFF in training, however, it falls short, as PFAS-containing AFFF can still be used if containment measures are taken. Washington state and North Carolina are two states that have enacted legislation to take this suggestion one step further by eliminating the use of PFAS-containing AFFF in all training exercises, regardless of containment measures. As military fire training and firefighting training occurs in a controlled environment, there is virtually no risk associated. Therefore, the surfactant does not need to be as effective and can be fluorine-free. Due to the lack of risk, the state legislation enacted by Washington state and North Carolina should be used as a model for establishing similar legislation on a federal level.

A "No PFAS in Firefighting Training Foam" legislation should be introduced, in which the use of AFFF for non-emergency measures should be eliminated, regardless of containment measures, to reduce any risk of contamination. Non-emergency measures include the firefighting training exercises performed by military training sites and fire departments. Emergency measures are classified as any oil-based fire, alternatively known as a liquid pool fire, that unexpectedly breaks out and has the potential to cause great structural damage, environmental harm, and injuries/casualties. The bill could be introduced by the House Committee on Armed Services, House Committee on Veterans' Affairs, House Committee on Science, Space, and Technology, and/or the House Committee on Transportation and Infrastructure.

IMPLEMENTATION

With the proposed complete ban of fluorine-containing foams for fire training purposes, a substitution will need to be used in place of fluorine-containing foam. The main purpose of fire training exercises is to ensure the trainees use proper technique. The training foams for liquid pool fires should have similar characteristics to previously used fluorinated foams. There are multiple types of fluorine-free foams. Several alternatives are listed below.

- GreenFire Fighting Foam (GFFF) [68]
- Fomtec Fluorine Free Foam [69]
- SOLBERG VERSAGARD AS-100 [70]

Appropriate proportioning devices must be used to dispense the foams and effectively extinguish oil-based Class B fires. For finding foams for training purposes, foams do not need to meet the military specifications. However, even for these alternatives, containment methods should be set up to reduce foam runoff and entrance of possible contaminants into the water system. Containment systems include pre-installed water barriers, permanent retention basins, fire-fighting water barriers, hoods and sealing pads, and mobile storage tanks [71]. By containing the foam-containing water through HAZMAT methods (i.e. damming, diking, diversion, and retainment), the potentially contaminated water can be transported to a proper water treatment facility.

By enacting "No PFAS in Firefighting Training Foam" legislation, there would be less PFAS use in firefighting sites and departments. PFAS exposure would be reduced for firefighters, military personnel, and trainees. Additionally, AFFF producers will see profit losses, as many PFAS-containing AFFF are used in training exercises [72]. Currently qualified product distributors for fluorinated foams that meet MilSpec MIL-F-24385F are listed in the appendix (see Appendix Table 4) [73].

The majority of companies with qualified fluorinated foam products for Class B fire military specifications have fluorine-free substitutions that could be used to replace the loss of revenue from AFFF sales. Research and development into fluorine-free foams is highly recommended for companies that do not have alternatives to PFAS-containing foams.

5.3.2 Replacement of All Fluorinated Foams with Fluorine-Free Foams

In the next five years, fluorinated firefighting foams should be phased out and replaced with fluorine-free foam. Although NDAA 2020 prohibits "uncontrolled release" of fluorinated AFFF at military sites starting on October 1, 2024, it grants an exception for emergency use [67][51]. By 2027, there should no longer be an exception for the "uncontrolled release" of fluorinated AFFF in emergency use, and only non-fluorinated foams should be allowed. **A portfolio of safe and effective fluorine-free foams should be developed.** Recent legislation in the Federal Aviation Act of 2020 has allowed fluorine-free foams to extinguish live fires at FAA-certified airports [74]. With this legislation, airports can explore various fluorine-free options and progressively build a portfolio.

The safe and effective fluorine-free foams portfolio should be developed and validated by December 2024. This identification should be spearheaded and sponsored by the Naval Research Laboratory and the DOD's Strategic Environmental Research and Development Program [75]. Once effective fluorine-free candidates are identified, the candidates can be tested at FAA-certified airports with varying fuel fires starting in October 2024. Similar to the initial fluorine-free foam use at the London Heathrow airport in 2015, select airports should switch to solely using fluorine-free foams and observe effectiveness over a 15-month evaluation period [76]. In this period, all use of fluorine-free foams, both emergency and training use, should be monitored and recorded. If a candidate shows early signs of ineffectiveness, the candidate will be removed for safety assurance. After these 15 months, the Department of Defense and the Naval Research Laboratory should examine the evaluation reports and approve or disapprove of a fluorine-free candidate. As more fluorine-free airports are tested, and a basis for extinguishing liquid pool fires is set, the legislation should extend to military sites, petrochemical plants, and fire training facilities.

Potential Obstacles

A current obstacle with replacing the PFAS-containing AFFF is that fire extinguishing agents for Class B fires must meet specifications from MIL-F-24385F. In the United States thus far, only fluorinated foams have met all the specifications contained in MIL-F-24385F. Work is being done to establish fluorine-free foams that have the same effectiveness as fluorinated aqueous film-forming foams. Although non-fluorinated foams have yet to meet every specification within the MilSpec, multiple alternatives have been tested for and have met the *fire extinction metrics* of MIL-F-24385F. NRL researcher Katherine Hinnant has done exploratory research with fluorine-free foams. Two of these foams, 3:2 Cap/Gluc and Cap, have been able to fully extinguish a live "pool fire" in under 30 seconds [77][78]. However, these foams do not meet all the standards listed in MIL-F-24385F, as each foam must have specific liquid properties.

It is recommended that the military specifications for MIL-F-24385F are reevaluated and potentially amended to expand the range of liquid properties. The amendment should shift focus to fire extinction metrics. For example, if a suppressant can extinguish a smaller 28 ft^2 pool fire within 30 seconds, in addition to meeting burnback requirements, the foam meets specifications. Through amending the MilSpec, the door opens for equally effective, fluorine-free foams to meet standards and be used in emergency live pool fires.

In the Interim

Until an equally effective fire extinguishing agent is found for Class B Fires, AFFF should continue to be used for all emergency fires to ensure safety and decrease risk. For live situations with emergency fires, containment setup may not be possible, as safety and extinguishing the fire is the top priority. If possible, the runoff and remaining firefighting agents should be cleaned up immediately after the situation is deemed safe. All sites with fire training should have an in-house wastewater center and water containment system to deal with AFFF used in live, emergency fires. By containing the PFAS-containing water, the contaminated or potentially contaminated water can be transported to a proper water treatment facility.

5.4 PFAS in the Water Supply

5.4.1 Safe Drinking Water - Water Treatment Plants

The EPA's National Primary Drinking Water Regulations does not currently regulate PFAs [33]. If primary standards are set in the immediate future, the standards are still not enforceable until three years after the EPA's finalization [36]. Congressional legislation for the enforcement of water treatment systems should happen immediately to mitigate future negative health and environmental effects. The legislation should take place as follows.

As previously mentioned, public water systems treat water and provide potable drinking water to communities of varying sizes. Current water treatment plants for community water systems require several steps for treatment and purification, as outlined in Section 3.3. These water treatment plants are useful for removing microorganisms, particles, and other enforced contaminants from drinking water; however, many are ineffective in removing PFAS from drinking water.

Subsequently, there is a great need to equip public water systems with effective PFAS treatment technologies. To tackle this effort, first, an official portfolio of acceptable and effective treatment methods for PFAS removal should be established by the EPA. Studied removal technologies that could be included in the portfolio are listed below.

Example: PFAS Removal Technologies Portfolio

Granular Activated Carbon

Granular activated carbon (GAC) is a highly porous material that can be used to filter out PFAS in water through adsorption. Adsorption is the movement of a substance from a gas or liquid phase to the surface of a solid [79][80]. GAC consists of various high-carbon organic materials, such as coal and wood [81]. GAC is effective at treating and removing long-chain

PFAS, such as PFOA and PFOS. However, GAC is not as effective at removing shorter chain PFAS from water [82]. After the PFAS compounds are adsorbed to the binding sites and the GAC is spent, the filter beds must be properly disposed of through either incineration or other methods. Despite these drawbacks, GAC removes many of the listed enforced contaminants, making it an attractive option for water systems due to its versatility [83].

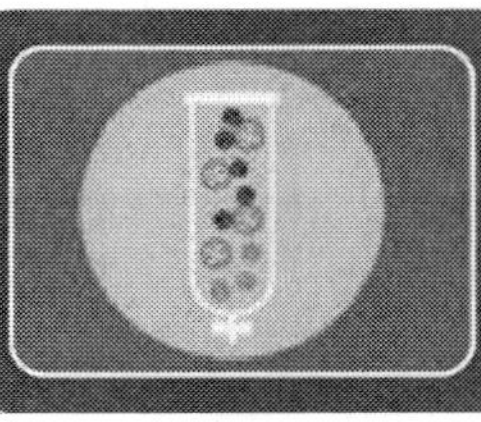

Ion exchange resins

Ion exchange resins consist of beads made up of highly porous material that is insoluble in water, acid, and base solutions. Ion exchange resins remove PFAS from solution through the use of anion exchange. The positively charged resin attracts the negatively PFAS compounds, causing them to bind to the resin [84]. There are two resin types: single-use and regenerable resins [84]. Single-use resins have been found to be more effective in the treatment of low-concentrations of PFAS, while regenerable resins are more effective in the treatment of high-concentrations of PFAS. Single-use resins are more applicable to drinking water treatment systems [82]. Although ion exchange resins are generally **more effective** in removing PFAS than GAC, they are also more costly than GAC, signifying a great financial burden for water systems [81].

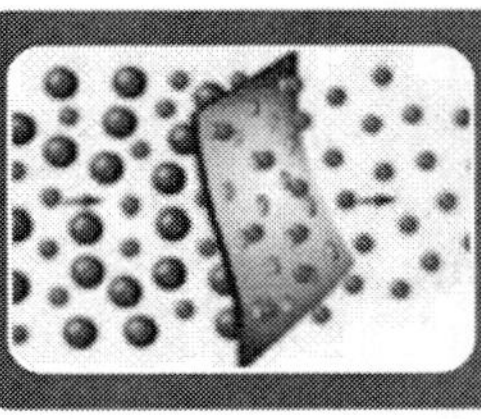

Reverse osmosis

In reverse osmosis, PFAS-contaminated water is pushed through a semipermeable membrane through the application of pressure [82]. Reverse osmosis is more than 90 percent effective at removing PFAS from water [81]. One consideration is that reverse osmosis filters sodium chloride from water, while other treatment techniques, such as nanofiltration, retain essential minerals [81]. Multiple technical challenges are associated with reverse osmosis, including membrane fouling, temperature parameters, and feedwater quality. Although effective, reverse osmosis requires a large amount of energy to power high-pressure pumps, making it a costly and energy-intensive option [82][84]. Reverse osmosis systems are commonly used in household purification systems, signifying a potential opportunity for PFAS-contaminated water purification for those communities that do not have adequate water infrastructure [80].

END OF EXAMPLE PORTFOLIO

Public water systems can choose to implement one of the approved treatment methods listed in the portfolio based on the resources and funds available. Due to the cost of technology implementation, **the implementation should only be enforced for public water systems that have at least one of the following classifications:**

a) **have already been impacted by measurable PFAS contamination, such as PFOA, PFOS, and PFHxS**
b) **are within the vicinity of known PFAS sources, such as military training bases and industrial facilities**

Additional Action: System Challenges

Although some public water systems have the resources and infrastructure to integrate PFAS-removing technologies, other small water systems can face multiple issues, such as aging infrastructure, lack of expertise, lack of financial resources, and lack of resources [85]. For these systems that lack the appropriate infrastructure and resources, alternative action should be taken to ensure safe drinking water. In these scenarios, individual water filtration systems and/or bottled water should be provided to the households within the affected area.

Individual water filtration systems should contain charcoal filters (granular activated carbon) or reverse osmosis filters. Due to the water inefficiency, reverse osmosis filters should only be used for one water source, for example, a kitchen sink or similar water dispenser. GAC systems should be used for entire households [86].

Funding

For public water systems where industry pollution is responsible for the PFAS contamination, the legislation should require that responsible parties are held accountable for funding these filtration systems and/or bottled water. To fund removal technologies for a state's water treatment plants, taxes should be increased for fluorochemical companies in the state. The increased tax income from these companies should be used to assist water utilities in equipping their treatment plants with adequate technologies. For communities that have utilities with insufficient resources and infrastructure, the tax revenue will go to ensuring safe water to communities through the purchase of bottled water and effective home filters. Additionally, loans can be taken from the Drinking Water State Revolving Fund (DWSRF) to more imminently fund improved systems and provide safe drinking water [87].

In the past, similar statewide legislation has been enacted to give potable water to those whose water sources have been contaminated by PFAS. In North Carolina, a prominent PFAS issue has been the GenX contamination of North Carolina's Cape Fear River by the Chemours Fayetteville works plant. In 2017, 19 households were found to have concentrations of GenX, a short-chain PFAS, above the state's provisional health goal. To ensure safe drinking water, state officials mandated that Chemours give bottled water to the affected households [88]. Considering this issue, representative Deb Butler of New Hanover County, NC, recently spearheaded water treatment through a new bill, H444, that would hold industry PFAS contaminators responsible for water treatment [89]. The bill was introduced to ensure safe drinking water supplies for those affected by GenX and specified PFAS contamination [90]. On a federal level, measures like these should be taken to ensure safe drinking water across the nation. If federal action is not immediately taken, in the interim, similar bills should be introduced and enacted in state legislatures for states that have high PFAS concentrations.

Future technologies & Challenges

PFAS Removal

Current technologies can remove PFAS, however, are limited in the amount and type of PFAS that can be removed. Additionally, all methods, although varying in price, are cost intensive. There is space for future research and development into purification technologies for PFAS removal. Funding should be appropriated to research institutions and universities for technology development to remove existing and emerging PFAS.

PFAS Disposal

All three methods can adsorb the PFAS and remove it from the water, but these materials still contain the PFAS. Further destruction technologies should be developed. Further, once the PFAS are removed from water, the materials with PFAS should be dealt with. The current destruction method is incineration of PFAS, where the PFAS materials are incinerated at extremely high temperatures and theoretically dissociated into the basic elements (i.e. carbon, fluorine, etc.) [84]. However, more research should be conducted to investigate if the PFAS are truly incinerated, or if they are merely degraded to shorter-chain fluorinated compounds. Additionally, research should be done on various destruction technologies so that existent PFAS can be effectively removed from the environment and waste streams.

Finally, as more research is conducted and more effective technologies are developed, grants should be given to affected water treatment facilities. These grants will enable water treatment facilities to equip facilities with effective remediation technologies and further remove PFAS from drinking water.

5.5 "Nothing but the Essentials" PFAS Ban

The ultimate solution for removing PFAS from drinking water is eliminating the use of PFAS within the industrial sector, and subsequently, the consumer market. By banning all PFAS, the point source of contamination would be removed, mitigating new contamination that could occur in public water systems. The ban would consist of a large-scale ban on PFAS industrial use, whether it is in producing PFAS precursor chemicals, manufacturing products with the PFAS surfactant properties, and using PFAS in production-related activities. A ban on all PFAS would serve as a preventative measure, focused on mitigating potential future risks based on negative health effects of legacy perfluorochemicals. Removing all PFAS from the market and production is a complex issue that cannot be taken lightly. Deliberate and carefully considered action should be taken to tackle a PFAS ban strategically.

In May 2021, the EU's European Chemical Agency (ECHA) published an update for the Chemical Sustainability Strategy (CSS), which calls for an eventual ban on all PFAS unless deemed to be essential [91] [92]. **The United States should follow the EU's strategy and enact a legislation that bans PFAS within the next ten years, by 2032**. The legislative plan should be developed within the House Committee on Energy and Commerce to draft a "Nothing but the Essentials" ban for PFAS. Although immediate action is needed to eliminate future drinking water contamination, time should be given so that actions are deliberately investigated to prevent potential negative outcomes.

In order to eliminate PFAS from the supply chain, substitutes for PFAS should be implemented. PFAS are found in multiple consumer products, including fast food wrappers, water-resistant coatings, and cleaning products, due to their physical properties. PFAS are desirable in consumer and safety products due to their numerous characteristics, listed below [11].

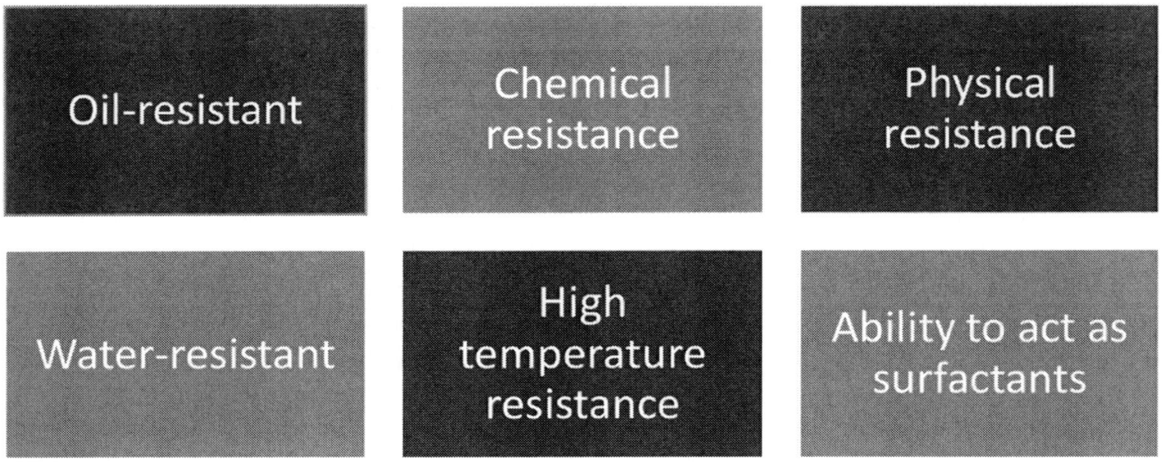

Substitutions for PFAS need to consider the aforementioned characteristics. There should be various types of substitutions for each characteristic. To supply products that have nearly identical properties, adequate appropriations should be put towards developing new chemicals that can imitate the key properties of PFAS.

Additionally, the search for and implementation of new chemicals that do not contain polyfluorinated carbons contains a challenge within itself: the potential toxicological and health effects associated with the new chemicals. There are several main issues associated with PFAS. Listed below are the undesirable characteristics associated with PFAS [11].

- Persistent in the environment after partial degradation
- Mobility in the environment (multiple mediums: air, water, soil)
- Bioaccumulative
- Toxic: associated with multiple health issues.

Substitutions for PFAS should not have these undesirable, negative traits. Past short-term solutions were focused on substituting the primary long-chain compounds of focus, PFOA and PFOS, for short-chain perfluorinated compounds. Due to having shorter half-lives, these shorter-chain replacement compounds were considered to not have the negative effects associated with PFOA and PFOS. However, research has shown that short-chain PFAS compounds can also be bioaccumulative, persistent, and toxic [93].

A past example of an adverse PFAS substitution was the replacement of PFOA with GenX. As a result of the PFOA phase-out, the short-chain GenX was used as a replacement for making Teflon-like materials. Later studies showed that GenX was bioaccumulative, persistent, and mobile in the environment. Additionally, recent animal laboratory studies have shown GenX is associated with health effects such as liver cancer, pancreatic cancer, and testicular cancer [94]. Although no human health effects have been observed yet, these preliminary animal studies show reason for concern of future human health effects. This example shows the importance of thorough investigation for potential PFAS substitutions. Markedly, future substitutions must be screened to ensure that there are no adverse effects.

To ensure the proposed substitutions are not toxic, multiple health, toxicity, and safety studies must be conducted. Once substitutes are located, the bill must provide appropriations to various agencies, primarily the NIEHS, the National Toxicology Program, and the EPA.

6 Conclusion

The topic of PFAS in drinking water is nuanced and has many items to consider. The author has suggested the following measures be taken to reduce PFAS in drinking water:

1. Individual states should enforce maximum contaminant levels for measurable PFAS in their drinking water on both an immediate and long-term basis.
2. The waste streams of chemical plants involved with PFAS must be heavily regulated.
3. There should be more regulation focused on reducing aqueous film-forming foam in drinking water, including the phase out of PFAS-containing foam and development of a comprehensive portfolio of effective fluorine-free foams.
4. To ensure safe drinking water, water treatment plants should be equipped with adequate PFAS treatment technologies. For areas with high concentrations of PFAS in water and inadequate water infrastructure, bottled water and household filters should be provided.
5. To significantly reduce the future presence of PFAS in drinking water, a large-scale "Nothing but the Essentials" ban on PFAS should be introduced. PFAS-free substitutes without the undesirable characteristics of PFAS – persistent, mobile, bioaccumulative, and toxic – should be developed and introduced into the market.

Implementation of the aforementioned suggestions could result in reduced PFAS in drinking water and safer drinking water. The author recognizes that there are many more aspects that need to be explored and hopes that the background and recommendations can spur further discussion on the matter.

7 Bibliography

[1] E. M. Sunderland, X. C. Hu, C. Dassuncao, A. K. Tokranov, C. C. Wagner, and J. G. Allen, "A Review of the Pathways of Human Exposure to Poly- and Perfluoroalkyl Substances (PFASs) and Present Understanding of Health Effects," *J. Expo. Sci. Environ. Epidemiol.*, vol. 29, no. 2, pp. 131–147, Mar. 2019, doi: 10.1038/s41370-018-0094-1.

[2] "The Lawyer Who Became DuPont's Worst Nightmare - The New York Times." https://www.nytimes.com/2016/01/10/magazine/the-lawyer-who-became-duponts-worst-nightmare.html (accessed Jul. 01, 2021).

[3] J. Mordock, "Taking on DuPont: Illnesses, deaths blamed on pollution from W. Va. plant," *The News Journal*. https://www.delawareonline.com/story/news/2016/04/01/dupont-illnesses-deaths-c8/81151346/ (accessed Jul. 01, 2021).

[4] O. US EPA, "What are PFCs and How Do They Relate to Per- and Polyfluoroalkyl Substances (PFASs)?," Jul. 06, 2016. https://www.epa.gov/pfas/what-are-pfcs-and-how-do-they-relate-and-polyfluoroalkyl-substances-pfass (accessed Aug. 04, 2021).

[5] "Perfluoroalkyl and Polyfluoroalkyl Substances (PFAS)," *National Institute of Environmental Health Sciences*. https://www.niehs.nih.gov/health/topics/agents/pfc/index.cfm (accessed Jul. 15, 2021).

[6] "PFAS_FamilyTree_EnvHealthPro-508.pdf." Accessed: Aug. 04, 2021. [Online]. Available: https://www.atsdr.cdc.gov/pfas/docs/PFAS_FamilyTree_EnvHealthPro-508.pdf

[7] "CLU-IN | Contaminants > Per- and Polyfluoroalkyl Substances (PFASs) > Overview." https://clu-in.org/contaminantfocus/default.focus/sec/Per-_and_Polyfluoroalkyl_Substances_(PFASs)/cat/Overview/ (accessed Aug. 04, 2021).

[8] "PFAS chemical exposure | ATSDR," Jun. 24, 2020. https://www.atsdr.cdc.gov/pfas/health-effects/exposure.html (accessed Jul. 22, 2021).

[9] "A guide to the PFAS found in our environment." https://cen.acs.org/sections/pfas.html (accessed Jul. 01, 2021).

[10] "The Differences Between Persistent & Non-Persistent Chemicals," *Sciencing*. https://sciencing.com/info-8751215-differences-between-persistent-nonpersistent-chemicals.html (accessed Aug. 04, 2021).

[11] "SWD_PFAS.pdf." Accessed: Jul. 19, 2021. [Online]. Available: https://ec.europa.eu/environment/pdf/chemicals/2020/10/SWD_PFAS.pdf

[12] "Definition of BIOACCUMULATION." https://www.merriam-webster.com/dictionary/bioaccumulation (accessed Aug. 04, 2021).

[13] "The History and Science Behind Teflon," *Product Release*, Dec. 14, 2015. https://www.product-release.com/product-release-news/history-science-behind-teflon/ (accessed Jul. 13, 2021).

[14] "The History of TeflonTM Fluoropolymers." https://www.teflon.com:443/en/news-events/history (accessed Jul. 13, 2021).

[15] B. E. Blake and S. E. Fenton, "Early life exposure to per- and polyfluoroalkyl substances (PFAS) and latent health outcomes: A review including the placenta as a target tissue and possible driver of peri- and postnatal effects," *Toxicology*, vol. 443, p. 152565, Oct. 2020, doi: 10.1016/j.tox.2020.152565.

[16] "History of PFAS and 3M." https://www.3m.com/3M/en_US/pfas-stewardship-us/pfas-history/ (accessed Jul. 13, 2021).

[17] X. C. Hu *et al.*, "Detection of Poly- and Perfluoroalkyl Substances (PFASs) in U.S. Drinking Water Linked to Industrial Sites, Military Fire Training Areas, and Wastewater Treatment Plants," *Environ. Sci. Technol. Lett.*, vol. 3, no. 10, pp. 344–350, Oct. 2016, doi: 10.1021/acs.estlett.6b00260.
[18] "3M, Lake Elmo reach $2.7M settlement in water suit," *MPR News*. https://www.mprnews.org/story/2019/05/21/3m-lake-elmo-reach-tentative-settlement-in-pfas-drinking-water-lawsuit (accessed Jul. 15, 2021).
[19] *City of Lake Elmo v. 3M Co.*, vol. 237. 2017, p. 877.
[20] J. Giesy and K. Kannan, "Global distribution of perfluorooctane sulfonate in wildlife.," *Environ. Sci. Technol.*, 2001, doi: 10.1021/ES001834K.
[21] O. US EPA, "Fact Sheet: 2010/2015 PFOA Stewardship Program," May 10, 2016. https://www.epa.gov/assessing-and-managing-chemicals-under-tsca/fact-sheet-20102015-pfoa-stewardship-program (accessed Aug. 29, 2021).
[22] A. M. Calafat, L.-Y. Wong, Z. Kuklenyik, J. A. Reidy, and L. L. Needham, "Polyfluoroalkyl Chemicals in the U.S. Population: Data from the National Health and Nutrition Examination Survey (NHANES) 2003–2004 and Comparisons with NHANES 1999–2000," *Environ. Health Perspect.*, vol. 115, no. 11, pp. 1596–1602, Nov. 2007, doi: 10.1289/ehp.10598.
[23] O. US EPA, "Basic Information on PFAS," *US EPA*, Mar. 30, 2016. https://www.epa.gov/pfas/basic-information-pfas (accessed Jun. 18, 2021).
[24] "Short-chain and long-chain PFAS show similar toxicity, US National Toxicology Program says," *Chemical & Engineering News*. https://cen.acs.org/environment/persistent-pollutants/Short-chain-long-chain-PFAS/97/i33 (accessed Aug. 29, 2021).
[25] O. US EPA, "Information about Public Water Systems," Sep. 21, 2015. https://www.epa.gov/dwreginfo/information-about-public-water-systems (accessed Aug. 04, 2021).
[26] "Providing Safe Drinking Water in America: 2013 National Public Water Systems Compliance Report," p. 97.
[27] "Public Water Systems | Drinking Water | Healthy Water | CDC," Mar. 30, 2021. https://www.cdc.gov/healthywater/drinking/public/index.html (accessed Jul. 15, 2021).
[28] "Water Treatment Process," *JCMariano*, Jul. 14, 2015. https://jcmariano.wordpress.com/2015/07/14/water-treatment-process/ (accessed Aug. 07, 2021).
[29] "7 Major Stages in Water Treatment Plant," *The Constructor*, Mar. 15, 2021. https://theconstructor.org/environmental-engg/water-supply/stages-water-treatment-plant/498706/ (accessed Aug. 05, 2021).
[30] Concerning Reality, *How Do Water Treatment Plants Work?* Accessed: Aug. 05, 2021. [Online Video]. Available: https://www.youtube.com/watch?v=0_ZcCqqpS2o
[31] "Treatment Process," *Denver Water*. https://www.denverwater.org/your-water/treatment-process (accessed Aug. 07, 2021).
[32] "Drinking Water Standards and Regulations | Public Water Systems | Drinking Water | Healthy Water | CDC," Nov. 03, 2020. https://www.cdc.gov/healthywater/drinking/public/regulations.html (accessed Aug. 29, 2021).

[33] O. US EPA, "National Primary Drinking Water Regulations," Nov. 30, 2015. https://www.epa.gov/ground-water-and-drinking-water/national-primary-drinking-water-regulations (accessed Jul. 23, 2021).
[34] O. US EPA, "Overview of the Safe Drinking Water Act," *US EPA*, Apr. 01, 2015. https://www.epa.gov/sdwa/overview-safe-drinking-water-act (accessed Jul. 02, 2021).
[35] "Understanding the Safe Drinking Water Act," p. 4.
[36] O. US EPA, "How EPA Regulates Drinking Water Contaminants," Sep. 01, 2015. https://www.epa.gov/sdwa/how-epa-regulates-drinking-water-contaminants (accessed Aug. 14, 2021).
[37] O. US EPA, "Types of Drinking Water Contaminants," Jun. 23, 2014. https://www.epa.gov/ccl/types-drinking-water-contaminants (accessed Aug. 29, 2021).
[38] M. Tiemann and E. H. Humphreys, "Regulating Drinking Water Contaminants: EPA PFAS Actions," p. 3.
[39] "Per- and Polyfluoroalkyl Substances (PFAS)." https://ntp.niehs.nih.gov/whatwestudy/topics/pfas/index.html (accessed Jul. 02, 2021).
[40] T. A. Slotkin, E. A. MacKillop, R. L. Melnick, K. A. Thayer, and F. J. Seidler, "Developmental neurotoxicity of perfluorinated chemicals modeled in vitro," *Environ. Health Perspect.*, vol. 116, no. 6, pp. 716–722, Jun. 2008, doi: 10.1289/ehp.11253.
[41] S. J. Frisbee *et al.*, "The C8 Health Project: Design, Methods, and Participants," *Environ. Health Perspect.*, vol. 117, no. 12, pp. 1873–1882, Dec. 2009, doi: 10.1289/ehp.0800379.
[42] "Potential health effects of PFAS chemicals | ATSDR," Jun. 24, 2020. https://www.atsdr.cdc.gov/pfas/health-effects/index.html (accessed Jun. 18, 2021).
[43] "PFAS Overview and Prevalence.pdf." Accessed: Jul. 14, 2021. [Online]. Available: https://www.awwa.org/Portals/0/AWWA/ETS/Resources/PFAS%20Overview%20and%20Prevalence.pdf?ver=2020-09-22-072340-407
[44] "2.6 PFAS Releases to the Environment – PFAS — Per- and Polyfluoroalkyl Substances." https://pfas-1.itrcweb.org/2-6-pfas-releases-to-the-environment/ (accessed Jul. 10, 2021).
[45] "Smokestack - Energy Education." https://energyeducation.ca/encyclopedia/Smokestack (accessed Jul. 15, 2021).
[46] T. Stoiber, S. Evans, and O. V. Naidenko, "Disposal of products and materials containing per- and polyfluoroalkyl substances (PFAS): A cyclical problem," *Chemosphere*, vol. 260, p. 127659, Dec. 2020, doi: 10.1016/j.chemosphere.2020.127659.
[47] "Leachate is the Driving Force for PFAS Sequestration in Landfills," *Waste Advantage Magazine*, Nov. 02, 2020. https://wasteadvantagemag.com/leachate-is-the-driving-force-for-pfas-sequestration-in-landfills/ (accessed Jul. 13, 2021).
[48] "Landfill Leachate - an overview | ScienceDirect Topics." https://www.sciencedirect.com/topics/earth-and-planetary-sciences/landfill-leachate (accessed Jul. 13, 2021).
[49] N. M. Robey, B. F. da Silva, M. D. Annable, T. G. Townsend, and J. A. Bowden, "Concentrating Per- and Polyfluoroalkyl Substances (PFAS) in Municipal Solid Waste Landfill Leachate Using Foam Separation," *Environ. Sci. Technol.*, vol. 54, no. 19, pp. 12550–12559, Oct. 2020, doi: 10.1021/acs.est.0c01266.
[50] "PFAS_Fact_Sheet_History_and_Use_April2020.pdf." Accessed: Jul. 13, 2021. [Online]. Available: https://pfas-1.itrcweb.org/fact_sheets_page/PFAS_Fact_Sheet_History_and_Use_April2020.pdf

[51] "Aqueous Film-Forming Foam (AFFF).pdf." Accessed: Jul. 11, 2021. [Online]. Available: https://pfas-1.itrcweb.org/wp-content/uploads/2020/10/afff_508_093020.pdf
[52] "MIL-F-24385F: Fire Extinguishing Agent, Aqueous Film-Forming Foam (AFFF) Liquid Concentrate, for Fresh and Sea Water."
[53] B. J. Place and J. A. Field, "Identification of Novel Fluorochemicals in Aqueous Film-Forming Foams (AFFF) Used by the US Military," *Environ. Sci. Technol.*, vol. 46, no. 13, pp. 7120–7127, Jul. 2012, doi: 10.1021/es301465n.
[54] "AC_150_5210-6D.pdf." Accessed: Jul. 11, 2021. [Online]. Available: https://www.faa.gov/documentLibrary/media/Advisory_Circular/AC_150_5210-6D.pdf
[55] O. US EPA, "Drinking Water Health Advisories for PFOA and PFOS," May 05, 2016. https://www.epa.gov/ground-water-and-drinking-water/drinking-water-health-advisories-pfoa-and-pfos (accessed Jul. 19, 2021).
[56] O. US EPA, "Third Unregulated Contaminant Monitoring Rule," *US EPA*, Sep. 01, 2015. https://www.epa.gov/dwucmr/third-unregulated-contaminant-monitoring-rule (accessed Jul. 02, 2021).
[57] O. US EPA, "Fifth Unregulated Contaminant Monitoring Rule," Jan. 11, 2021. https://www.epa.gov/dwucmr/fifth-unregulated-contaminant-monitoring-rule (accessed Jul. 19, 2021).
[58] "pfas_action_plan_feb2020.pdf." Accessed: Jul. 15, 2021. [Online]. Available: https://www.epa.gov/sites/production/files/2020-01/documents/pfas_action_plan_feb2020.pdf
[59] M. Simic, "The Third Unregulated Contaminant Monitoring Rule (UCMR 3): Data Summary, January 2017," p. 12, 2017.
[60] O. US EPA, "Occurrence Data for the Unregulated Contaminant Monitoring Rule," Sep. 01, 2015. https://www.epa.gov/dwucmr/occurrence-data-unregulated-contaminant-monitoring-rule (accessed Jul. 19, 2021).
[61] W. Staff, "Chemours' thermal oxidizer destroys PFAS air emissions above state-mandated efficiency level," *https://www.wect.com*. https://www.wect.com/2020/03/30/chemours-thermal-oxidizer-destroys-pfas-air-emissions-above-state-mandated-efficiency-level/ (accessed Jul. 23, 2021).
[62] R. Riley, "Chemours Fayetteville site's air emission equipment operational," *The Fayetteville Observer*. https://www.fayobserver.com/news/20191227/chemours-fayetteville-sitersquos-air-emission-equipment-operational (accessed Jul. 23, 2021).
[63] O. US EPA, "Municipal Solid Waste Landfills," Mar. 24, 2016. https://www.epa.gov/landfills/municipal-solid-waste-landfills (accessed Jul. 26, 2021).
[64] "pfas_methods-sampling_tech_brief_7jan2020-update.pdf." Accessed: Aug. 26, 2021. [Online]. Available: https://www.epa.gov/sites/default/files/2020-01/documents/pfas_methods-sampling_tech_brief_7jan2020-update.pdf
[65] O. US EPA, "Learn the Basics of Hazardous Waste," Nov. 25, 2015. https://www.epa.gov/hw/learn-basics-hazardous-waste (accessed Aug. 30, 2021).
[66] "Definitions of Hazwoper terms including immediate, incidental & uncontrolled release. | Occupational Safety and Health Administration." https://www.osha.gov/laws-regs/standardinterpretations/1990-11-14-0 (accessed Jul. 17, 2021).
[67] J. M. Inhofe, "Text - S.1790 - 116th Congress (2019-2020): National Defense Authorization Act for Fiscal Year 2020," Dec. 20, 2019. https://www.congress.gov/bill/116th-congress/senate-bill/1790/text (accessed Jul. 13, 2021).

[68] "Fluorine Free Foam Replacement for AFFF | Greenfire®." https://gogreenfire.com/fluorine-free-foam-replacement-for-afff/?web=1&wdLOR=cD857696A-CEB8-FE43-9399-C3DEDD43EE99 (accessed Aug. 28, 2021).
[69] "43f81add-b874-42f1-81f3-2d41ca8dad5b.pdf." Accessed: Aug. 28, 2021. [Online]. Available: https://www.fire-protection.com.au/CMSTemplates/files/fireprotection.com.au/43/43f81add-b874-42f1-81f3-2d41ca8dad5b.pdf
[70] "SOLBERG® VERSAGARDTM AS-100 FLUORINE-FREE," *Perimeter Solutions*. https://www.perimeter-solutions.com/class-b-foam/solberg-versagard-as-100-fluorine-free/ (accessed Jul. 11, 2021).
[71] "d5b24e2a-d027-0168-cdd8-f723c675fa98.pdf." Accessed: Jul. 17, 2021. [Online]. Available: https://echa.europa.eu/documents/10162/28801697/pfas_flourine-free_alternatives_fire_fighting_en.pdf/d5b24e2a-d027-0168-cdd8-f723c675fa98
[72] "House OKs Limits On Firefighting Foam With PFAS | Coastal Review," May 11, 2021. https://coastalreview.org/2021/05/house-passes-new-rules-to-restrict-firefighting-foam/ (accessed Jul. 20, 2021).
[73] "QPD Search." https://qpldocs.dla.mil/search/parts.aspx?qpl=1910&param=QPL-24385&type=256 (accessed Jul. 18, 2021).
[74] B. Guthrie, "Text - H.R.302 - 115th Congress (2017-2018): FAA Reauthorization Act of 2018," Oct. 05, 2018. https://www.congress.gov/bill/115th-congress/house-bill/302/text (accessed Jul. 12, 2021).
[75] "5 Things to Know About DOD," *U.S. Department of Defense*. https://www.defense.gov/Explore/News/Article/Article/1953510/5-things-to-know-about-dods-research-on-fluorine-free-firefighting-foam/ (accessed Jul. 17, 2021).
[76] "What is the price of fire safety?," *Chemical & Engineering News*. https://cen.acs.org/business/specialty-chemicals/price-fire-safety/97/i2 (accessed Jul. 10, 2021).
[77] K. Hinnant, S. Giles, A. Snow, J. Farley, J. Fleming, and R. Ananth, "An Analytically Defined Fire-Suppressing Foam Formulation for Evaluation of Fluorosurfactant Replacement," *J. Surfactants Deterg.*, vol. 21, Jul. 2018, doi: 10.1002/jsde.12166.
[78] K. M. Hinnant, S. L. Giles, E. P. Smith, A. W. Snow, and R. Ananth, "Characterizing the Role of Fluorocarbon and Hydrocarbon Surfactants in Firefighting-Foam Formulations for Fire-Suppression," *Fire Technol.*, vol. 56, no. 4, pp. 1413–1441, Jul. 2020, doi: 10.1007/s10694-019-00932-7.
[79] "Adsorption - an overview | ScienceDirect Topics." https://www.sciencedirect.com/topics/earth-and-planetary-sciences/adsorption (accessed Aug. 30, 2021).
[80] "What is Adsorption?," *The International Adsorption Society*. https://www.int-ads-soc.org/what-is-adsorption/ (accessed Aug. 30, 2021).
[81] O. US EPA, "Reducing PFAS in Drinking Water with Treatment Technologies," Aug. 23, 2018. https://www.epa.gov/sciencematters/reducing-pfas-drinking-water-treatment-technologies (accessed Aug. 14, 2021).
[82] "12 Treatment Technologies – PFAS — Per- and Polyfluoroalkyl Substances." https://pfas-1.itrcweb.org/12-treatment-technologies/ (accessed Jul. 23, 2021).
[83] "CFR-2007-title40-vol22-part142.pdf."

[84] "'Forever chemicals' no more? These technologies aim to destroy PFAS in water," *Chemical & Engineering News*. https://cen.acs.org/environment/persistent-pollutants/Forever-chemicals-technologies-aim-destroy/97/i12 (accessed Jul. 27, 2021).
[85] O. US EPA, "Small Drinking Water Systems Research," Oct. 19, 2015. https://www.epa.gov/water-research/small-drinking-water-systems-research (accessed Aug. 14, 2021).
[86] "In-Home Water Filtration Options – NH PFAS Investigation." https://www4.des.state.nh.us/nh-pfas-investigation/?page_id=171 (accessed Aug. 25, 2021).
[87] O. US EPA, "How the Drinking Water State Revolving Fund Works," Sep. 29, 2015. https://www.epa.gov/dwsrf/how-drinking-water-state-revolving-fund-works (accessed Aug. 29, 2021).
[88] "NC DEQ: State directs Chemours to provide bottled water to eight more well owners after latest preliminary tests for GenX." https://deq.nc.gov/state-directs-chemours-provide-bottled-water-eight-more-well-owners-after-latest-preliminary-tests (accessed Jul. 23, 2021).
[89] F. Weller, "PAY UP: Lawmaker wants Chemours to buy equipment to rid drinking water of toxic chemicals," *https://www.wect.com*. https://www.wect.com/2021/03/27/pay-up-lawmaker-wants-chemours-buy-equipment-rid-drinking-water-toxic-chemicals/ (accessed Jul. 23, 2021).
[90] "H444v1.pdf." Accessed: Jul. 23, 2021. [Online]. Available: https://www.ncleg.gov/Sessions/2021/Bills/House/PDF/H444v1.pdf
[91] "European Commission Update: The Future REACH Revision and PFAS Restriction," *JD Supra*. https://www.jdsupra.com/legalnews/european-commission-update-the-future-5463718/ (accessed Jul. 27, 2021).
[92] "CSS-Analysis_Final.pdf." Accessed: Jul. 27, 2021. [Online]. Available: https://www.ciel.org/wp-content/uploads/2021/01/CSS-Analysis_Final.pdf
[93] M. I. Gomis, R. Vestergren, D. Borg, and I. T. Cousins, "Comparing the toxic potency in vivo of long-chain perfluoroalkyl acids and fluorinated alternatives," *Environ. Int.*, vol. 113, pp. 1–9, Apr. 2018, doi: 10.1016/j.envint.2018.01.011.
[94] "GenX factsheet FINAL 13Sep2017.pdf." Accessed: Jul. 19, 2021. [Online]. Available: https://files.nc.gov/ncdeq/GenX/GenX%20factsheet%20FINAL%2013Sep2017.pdf
[95] O. US EPA, "PFAS Analytical Methods Development and Sampling Research," Apr. 21, 2020. https://www.epa.gov/water-research/pfas-analytical-methods-development-and-sampling-research (accessed Jul. 23, 2021).
[96] "otm_45_semivolatile_pfas_1-13-21.pdf." Accessed: Jul. 23, 2021. [Online]. Available: https://www.epa.gov/sites/default/files/2021-01/documents/otm_45_semivolatile_pfas_1-13-21.pdf
[97] "method_0010_0.pdf." Accessed: Jul. 23, 2021. [Online]. Available: https://www.epa.gov/sites/default/files/2020-04/documents/method_0010_0.pdf
[98] "to-15r.pdf." Accessed: Jul. 23, 2021. [Online]. Available: https://www3.epa.gov/ttnamti1/files/ambient/airtox/to-15r.pdf
[99] "EPA-HQ-OPPT-2002-0051-0003_content.pdf."

8 Appendix

Table 1. Companies and sites involved in the EPA's 2010/2015 Stewardship Program [38]

Company	Location	Nearby Watershed
DuPont	Washington Works, West Virginia	Ohio River
DuPont	Richmond, Virginia	James River
DuPont	Parlin, New Jersey	Raritan River
The Chemours Company	Fayetteville, North Carolina	Cape Fear River
3M	Decatur, Alabama	Tennessee River
3M	Cottage Grove, Minnesota	Mississippi River
3M	Cordova, Illinois	Mississippi River
Arkema Inc	Calvert City, Kentucky	Tennessee River
Asashi Glass Fluoropolymers USA	Bayonne, New Jersey	Hudson River
BASF	McIntosh, Alabama	Tombigbee River
Clariant Corporation	Mt. Holly, North Carolina	Catawba River
Solvay Solexis	West Deptford, New Jersey	Delaware River
ICT Chemicals	Cartersville, Georgia	Etowah River
First Chemical Corporation (of The ChemOurs Company)	Pascagoula, Mississippi	Bayou Casotte

Table 2. PFAS detection data collected under UCMR 3 [95]

PFAS Name	Type	Length
Perfluorooctanesulfonic acid (PFOS)	Legacy (phased out)	Long - 8 carbon chain
Perfluorooctanoic acid (PFOA)	Legacy (phased out)	Long - 8 carbon chain
Perfluorononanoic acid (PFNA)	Legacy	Long - 9 carbon chain
Perfluorohexanesulfonic acid (PFHxS)	Legacy	Short - 6 carbon chain
Perfluoroheptanoic acid (PFHpA)	Legacy	Short - 7 carbon chain
Perfluorobutanesulfonic acid (PFBS)	Legacy and Replacement	Short - 4 carbon chain

Table 3. Approved analytical test methods for PFAS source emissions [96]

Test Method	Description
Other Test Method (OTM)-45	Measures emissions of 50 PFAS from vents and stacks [97]
SW-846 Test Method 0010: Modified Method 5 Sampling Train	Determines the Destruction and Removal Efficiency of incineration systems that remove semi/non-volatiles compounds [98]
Modified Method TO-15	Measures 4 volatile perfluorinated volatile compounds [73]

Table 4. Qualified Product List for Extinguishing Class B Fires [99]

Foam Distributor	Product	Fluorine-Free Substitute
National Foam, Inc.	AER-O-WATER 3EM-C6 AFFF AER-O-WATER 6EM-C6 AFFF	Yes
Perimeter Solutions	ARCTIC 3% MIL-SPEC AFFF PHOS-CHEK 3% AFFF MS ARCTIC 6% MIL-SPEC AFFF PHOS-CHEK 6% MILSPEC AFFF	Yes
Tyco Fire Products / Johnson Controls	ANSULITE AFC-3MS 3% AFFF CHEMGUARD C306-MS 3% AFFF ANSULITE AFC-6MS 6% AFFF CHEMGUARD C606-MS 6% AFFF	Yes
DAFO Fomtec AB	FOMTEC AFFF 3%M "SWE" FOMTEC AFFF 3%M "USA"	Yes
Buckeye Fire Equipment Company	PLATINUM PLUS C6 3% MS-AFFF PLATINUM PLUS C6 6% MS-AFFF	No
The Viking Corporation	VIKING AFFF3%M C6 "SWE" VIKING AFFF3%M C6 "USA"	No

Text 1. 3M Report – Waste of PFAS into Landfill

3M's "Fluorochemical Use, Distribution, and Release Overview" report from May 26, 1999 briefly discussed waste disposal of their perfluorooctanesulfonyl fluoride (POSF) production process biproducts [19]. During their manufacturing process, both byproducts and waste products that contained PFAS were created. While some byproducts could be recycled back into their process, other byproducts were released into internal wastewater treatment systems. Additionally, the treatment sludge waste products were either disposed of into landfills or applied to land. The effects of this type of disposal method could be seen through the contamination of drinking water in Lake Elmo, Minnesota. For over half a century, PFAS from 3M were dumped into various landfills in the area of Lake Elmo, such as the Oakdale Dump Lake Elmo's Washington Country Landfill. PFAS from the landfills had leached into the groundwater and therefore waste management facilities, contaminating a valuable drinking water source [19]. Due to this, the site is now regulated by the EPA and is a designated superfund site.

635296 CACS – Its Long History and Contributions to the Chemical Community

Abstract:

Chinese American Chemical Society (CACS), a non-political and non-profit professional organization, was founded by the late Dr. Jesse Hua 40 years ago during the 181st ACS National Meeting in Atlanta. I was privileged to be invited by Jesse Hwa to participate at its organization meeting and served as CACS Chair of the Board from 1997 to 2002. I helped the formation of the Great Lakes Chapter of CACS and have been serving as its board chair since its inauguration in 1997 to present.

The purpose of CACS is to encourage the advancement of chemistry and chemical engineering, to improve the qualification and occupational opportunities of its members, to facilitate professional contacts, and to promote interactions with other scientific communities. CACS is the only Chinese professional organization that is officially recognized by ACS and AIChE. Under the leadership of the current Chair of the Board, Dr. Marinda Wu, a former president of ACS, CACS and ACS signed the Chemistry Enterprise Partnership (CEP). The agreement is characterized by the shared commitment to cooperation in service to chemists, chemical engineers, students, educators, and professionals.

CACS has made valuable contributions to the chemical community. It holds technical forum and banquet at each of the national meeting of ACS and AIChE and through the wide ranged activities that are organized by its three local chapters. Looking toward the future, CACS will expand its activities to include more technical conferences. It also plans to have more cooperative programs with the various divisions of ACS and AIChE. During this talk, I will share some of my personal experience in serving CACS and perspectives toward its future growth.

635295 Chemical Looping Combustion, Gasification and Reforming: Perspectives in Commercialization Readiness

Abstract:

Chemical looping technology is a manifestation of the interplay among a broad spectrum of science and engineering subjects in connection to metal oxide physics, chemistry and reaction engineering, and particle science and technology. These subjects encompass three main components - Materials: metal oxide material synthesis, reactivity, redox mechanism recyclability, and physical strength; Reactors: flow pattern and stability, gas-solid contact mechanics, scaling rule; and Systems: process integration, intensification, and optimization. Such interplay is so complex that it has been over 100 years in which this technology has not been able to be commercially deployed.

However, with major advances made recently on the bottleneck areas in this technology development, the technology commercialization is now realistically possible and can be projected to take place in the near future. These advances include the successful development of chemically, physically robust metal oxide oxygen carriers that are cost-effective and sustainable to long-term redox reactor environment, and the successful employment of CO_2 and H_2O as partial substitute of carbonaceous feedstock for combustion, gasification and reforming applications thereby yielding CO_2negative chemical looping processes. These advances coupled with a novel reactor system design and operation give rise to a significant reduction in the capex compared to conventional approaches, when the chemical looping technology is used for carbonaceous feedstock conversions in the production of electricity, hydrogen, syngas, liquid fuels, and chemicals. This presentation will describe these advances that are established over the OSU moving-bed based chemical looping system platform. The presentation will also illustrate the rationales of its commercialization timeline with relevance to that for the fluidized bed-based chemical looping technology.

635294 Coupling CO_2 Reduction with Upgrading of Light Alkanes

Abstract:

The global CO_2 emission has increased its atmospheric concentration to 415 ppm in 2019, which has been linked to climate change and ocean acidification. In the meantime, a huge reserve (ca. 16.1 quadrillion cubic feet) of shale gas has been discovered globally. While the prominent component in shale gas is methane, other light alkanes (ethane, propane and butane) account for up to 15% hydrocarbons in shale basins and are typically underutilized. Therefore, it is highly desired to identify catalysts that promote the reactions of CO_2 and light alkanes. Using a combination of kinetic studies, *in situ* characterization and DFT calculations, we have identified several classes of selective catalysts that can break the C-C bonds to produce synthesis gas or the C-H bonds to produce olefins. We have also demonstrated the feasibility to further upgrade the syngas/olefin product mixture to value-added oxygenates and aromatics using cascade reactions and tandem reactors. In this presentation we will summarize recent advances and discuss challenge in coupling CO_2 reduction with the upgrading of light alkanes.

635293 Towards Platinum-Free Fuel Cells for Affordable Zero-Emission Vehicles

Abstract:

Green hydrogen from wind and solar electricity is necessary to decarbonize certain sectors of our economy that are inaccessible by renewable electricity and it has the potential to reduce more than 30% of the global carbon emission. For the transportation sector, green hydrogen can help decarbonization through its use in fuel cells. Green hydrogen refers to hydrogen that is produced by water electrolyzers powered by renewable electricity. For low temperature fuel cells and electrolyzers, polymer electrolytes play a critical role in controlling their cost, performance, and durability, and consequently their economic viability. In this presentation, I will focus on our recent work on hydroxide exchange membrane fuel cells (HEMFCs). More specifically I will show the roadmap we have developed for this technology, the progress we have made in developing one of the most stable membranes and some of the most active nonprecious metal catalysts. I will also try to answer the fundamental question: why are hydrogen oxidation reactions are slower in base than in acid for precious metal catalysts?

626587 The Influence of Geometry on the Mechanical Properties of Kirigami-Inspired Thin Films

Abstract:

Kirigami, the Japanese art of cutting shapes into sheets of paper, has found utility in materials science as a method to engineer a material's mechanical properties. By creating precise geometric cuts, kirigami-inspired films can be imbued with properties such as enhanced stretchability and conformability. Structural designs can be integrated with functional materials to form biodevices and surfaces for health monitoring. Computational methods to understand how kirigami patterns influence mechanics can be used for in-silico design of multifunctional materials with specific properties.

This study explores how cut geometries impact the stretching behavior of a thin film. A slit geometry was investigated to impart stretchability. The slits are modeled using an open-source kirigami design simulator GamIan. Finite-element analysis is used to visualize stretching behavior under strain. Various cut/slit parameters were studied using polyimide as a model substrate -e.g. cut width, cut length and cut spacing (by varying the number of cuts on the sheet). A 3D mesh of the designed thin films was simulated via a simple tensile test fixing one edge of the sheet, and pulling the other end over a number of steps. The elastic behavior is recorded via a stress-strain curve. The influence of each parameter was quantified individually and in combination. The stretch could also be visualized to better understand how the films behave under stress (**Figure 1**).

For instance, increasing the cut widths and decreasing cut spacings result in a steadily increasing and controlled elongation. Wider edges to the cuts, or minimizing the cut spacing allows the sheet to elongate without accumulating excessive stress in the structure. Instead, increasing stress is evenly distributed among the opening slits as the film is elongated. Thus, when designing a film device, decreasing cut spacing can be prioritized over increasing cut width to preserve material and ease of microfabrication. Polyimide films are widely used in thin film electronics as substrates for wearable devices. Our work further explored a silk fibroin biomaterial to show how this technique can be used for bio interfacing applications. This technique can be broadly applied to design complex geometries to induce specific properties for applications in healthcare and electronics.

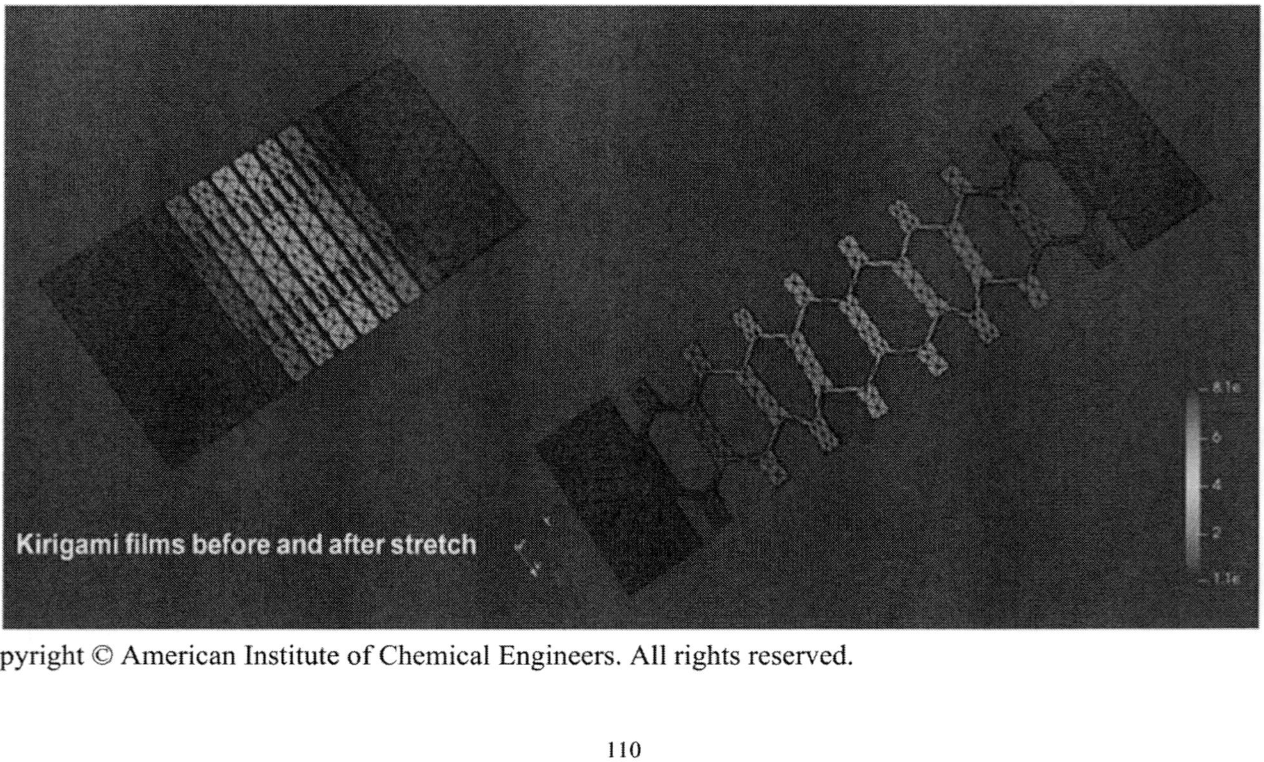

637411 The Hard Skills of Emotional Intelligence for Effective Engineering Leaders

Abstract:

High performing teams and leaders with emotional intelligence are more likely to achieve their goals. This workshop featuring Dr David Caruso of Yale's Center for Emotional Intelligence, focuses on the ability model of emotional intelligence and begins with a hands-on illustration of the four "hard" skills of EI, followed by demonstrations of the notion that "emotions are data" and the cognitive costs of emotion suppression. After a review of a brief case study, we then provide participants with a means to apply these skills to situations faced by leaders in engineering using the Emotional Intelligence Blueprint.

AUTHOR INDEX